State-Space Control Systems

The MATLAB®/Simulink® Approach

Synthesis Lectures on Controls and Mechatronics

Editors
Chaouki Abdallah, *University of New Mexico*
Mark Spong, *University of Texas at Dallas*

© Springer Nature Switzerland AG 2022

Reprint of original edition © Morgan & Claypool 2020

State-Space Control Systems: The MATLAB®/Simulink® Approach

Farzin Asadi

ISBN: 978-3-031-00704-0   paperback
ISBN: 978-3-031-01832-9   ebook
ISBN: 978-3-031-00085-0   hardcover

DOI 10.1007/978-3-031-01832-9

A Publication in the Springer series
SYNTHESIS LECTURES ON CONTROLS AND MECHATRONICS

Lecture #6
Series Editors: Chaouki Abdallah, *University of New Mexico*
                Mark Spong, *University of Texas at Dallas*
Series ISSN
Print 1939-0564   Electronic 1939-0572

State-Space Control Systems

The MATLAB®/Simulink® Approach

Farzin Asadi
Maltepe University, Istanbul, Turkey

SYNTHESIS LECTURES ON CONTROLS AND MECHATRONICS #6

ABSTRACT

These days, nearly all the engineering problem are solved with the aid of suitable computer packages. This book shows how MATLAB/Simulink could be used to solve state-space control problems.

In this book, it is assumed that you are familiar with the theory and concepts of state-space control, i.e., you took or you are taking a course on state-space control system and you read this book in order to learn how to solve state-space control problems with the aid of MATLAB/Simulink.

The book is composed of three chapters. Chapter 1 shows how a state-space mathematical model could be entered into the MATLAB/Simulink environment. Chapter 2 shows how a nonlinear system could be linearized around the desired opperating point with the aid of tools provided by MATLAB/Simulink. Finally, Chapter 3 shows how a state-space controller could be designed with the aid MATLAB and be tested with Simulink.

The book will be usefull for students and practical engineers who want to design a state-space control system.

KEYWORDS

control engineering, controller, controllability, feedback control, full state feedback, observability, sensor, Simulink®, state space, MATLAB®

In loving memory of my father, Moloud Asadi,
always on my mind, forever in my heart.

Farzin Asadi

Contents

Preface

Proportional-integral-derivative (PID) control uses feedback to detect errors between the desired system output and the actual system output and applies corrective commands to compensate for those errors.

Although PID control is the most common type of industrial controller, it does have limitations. First, PID control is generally not suitable for systems with multiple inputs and multiple outputs (MIMO), as the transfer functions and differential equations used to represent the system become overly complex when more then one input (or output) is involved. Second, PID control is based on constant parameters, so its effectiveness in controlling nonlinear systems is limited.

An alternative control method is state-space control. The key difference between PID control (aka "transfer control") and state-space control is that the state-space method takes into account the internal state of the system, through what are referred to as state variables. State variables are variables whose values evolve over time in a way that depends on the values they have at any given time and on the externally imposed values of input variables. Output variables' values depend on the values of the state variables.

These days, nearly all the engineering problem are solved with the aid of suitable computer packages. This book shows how MATLAB/Simulink could be used to solve state-space control problems. The book is composed of three chapters. Chapter 1 shows how a state-space mathematical model could be entered into the MATLAB/Simulink environment. Nearly all the systems in the real world are nonlinear. In order to use the linear techniques, first of all, a linear model of system must be obtained. Chapter 2 shows how a nonlinear system could be linearized around the desired opperating point with the aid of tools provided by MATLAB/Simulink. Chapter 3 shows how a state-space controller could be designed with the aid MATLAB and be tested with Simulink.

In this book, it is assumed that you are familiar with the state-space concepts, i.e., you took/taking a course on state-space control system and you want to learn how to solve the state-space control problems with the aid of MATLAB/Simulink. The book will be useful for practical engineers who also want to design a state-space controller.

Farzin Asadi
September 2020
farzinasadi@maltepe.edu.tr

CHAPTER 1

Entering the State-Space Model into MATLAB/Simulink Environment

1.1 INTRODUCTION

State-space representation is an important tool and is the preferred representation in many branches of control engineering. In this chapter, you will become familiar with the state-space representation and related commands and blocks in MATLAB and Simulink. The tools and techniques introduced in this chapter make the basis for other chapters of this book.

1.2 STATE-SPACE REPRESENTATION

A state-space representation for a linear time-invariant system has the general form

$$
\begin{aligned}
\dot{x}(t) &= Ax(t) + Bu(t) \\
y(t) &= Cx(t) + Du(t) \quad x(t_0) = x_0
\end{aligned}
\tag{1.1}
$$

in which $x(t)$ is the n-dimensional *state vector*

$$
x(t) = \begin{bmatrix} x_1(t) \\ x_2(t) \\ \vdots \\ x_n(t) \end{bmatrix}
\tag{1.2}
$$

whose n scalar components are called *state variables*. Similarly, the m-dimensional *input vector* and p-dimensional *output vector* are given, respectively, as

$$u(t) = \begin{bmatrix} u_1(t) \\ u_2(t) \\ \vdots \\ u_m(t) \end{bmatrix} \qquad y(t) = \begin{bmatrix} y_1(t) \\ y_2(t) \\ \vdots \\ y_p(t) \end{bmatrix}. \tag{1.3}$$

Since differentiation with respect to time of a time-varying vector quantities performed component-wise, the time-derivative on the left-hand side of Equation (1.1) represents

$$\dot{x}(t) = \begin{bmatrix} \dot{x}_1(t) \\ \dot{x}_2(t) \\ \vdots \\ \dot{x}_n(t) \end{bmatrix}. \tag{1.4}$$

Finally, for a specified initial time $t0$, the *initial state* $x(t0) = x0$ is a specified, constant, n-dimensional vector. The state vector $x(t)$ is composed of a minimum set of system variables that uniquely describes the future response of the system given the current state, the input, and the dynamic equations. The input vector $u(t)$ contains variables used to actuate the system, the output vector $y(t)$ contains the measurable quantities, and the state vector $x(t)$ contains internal system variables. Using the notational convention $M = [mij]$ to represent the matrix whose element in the ith row and jth column is mij, the coefficient matrices in Equation (1.1) can be specified via

$$A = \begin{bmatrix} a_{ij} \end{bmatrix} \quad B = \begin{bmatrix} b_{ij} \end{bmatrix} \quad C = \begin{bmatrix} c_{ij} \end{bmatrix} \quad D = [d_{ij}] \tag{1.5}$$

having dimensions $n \times n$, $n \times m$, $p \times n$, and $p \times m$, respectively. With these definitions in place, we see that the state Equation (1.1) is a compact representation of n scalar first-order ordinary differential equations, that is,

$$\dot{x}_i(t) = a_{i1}x_1(t) + a_{i2}x_2(t) + \cdots + a_{in}x_n(t) + b_{i1}u_1(t)$$
$$+ b_{i2}u_2(t) + \cdots + b_{im}u_m(t) \tag{1.6}$$

for $i = 1, 2, \ldots, n$, together with p scalar linear algebraic equations

$$y_j(t) = c_{j1}x_1(t) + c_{j2}x_2(t) + \cdots + c_{jn}x_n(t) + d_{j1}u_1(t)$$
$$+ d_{j2}u_2(t) + \cdots + d_{jm}u_m(t) \tag{1.7}$$

for $j = 1, 2, \ldots, p$. From this point on the vector notation (1.1) will be preferred over these scalar decompositions. The state-space description consists of the state differential equation

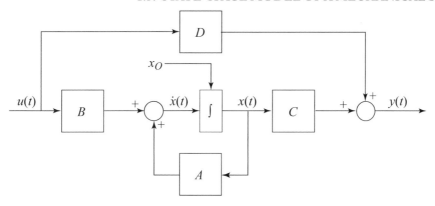

Figure 1.1: Block diagram for general multiple input, multiple-output, linear time-invariant system.

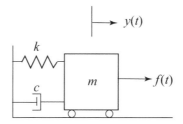

Figure 1.2: Translational mechanical system.

$\dot{x}(t) = Ax(t) + Bu(t)$ and the algebraic output equation $y(t) = Cx(t) + Du(t)$ from Equation (1.1).

Figure 1.1 shows the block diagram for the state-space representation of general multiple-input, multiple-output, linear time-invariant systems.

One motivation for the state-space formulation is to convert a coupled system of higher-order ordinary differential equations, for example, those representing the dynamics of a mechanical system, to a coupled set of first-order differential equations. In the single-input, single-output case, the state-space representation converts a single nth-order differential equation into a system of n-coupled first-order differential equations. In the multiple-input, multiple-output case, in which all equations are of the same order n, one can convert the system of k nth-order differential equations into a system of kn coupled first-order differential equations.

1.3   STATE-SPACE MODEL OF A MECHANICAL SYSTEM

Assume a mechanical system like the one shown in Fig. 1.2.

The free body diagram of this system is shown in Fig. 1.3.

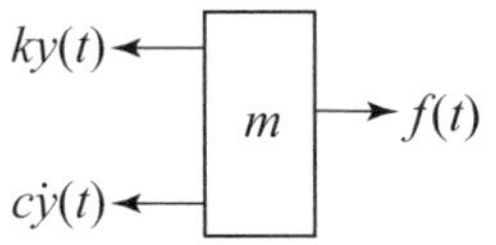

Figure 1.3: Free body diagram of mechanical system shown in Fig. 1.1.

Using Newton second law:

$$m\ddot{y}(t) + c\dot{y}(t) + ky(t) = f(t).$$ (1.8)

Since this is a second-order differential equation, the state vector need to be 2×1, i.e., $x(t) = \begin{bmatrix} x_1(t) \\ x_2(t) \end{bmatrix}$. Let select $x_1(t) = y(t)$ and $x_2(t) = \dot{y}(t) = \dot{x}_1(t)$. Then,

$$\dot{y}(t) = x_2(t)$$
$$\ddot{y}(t) = \dot{x}_2(t).$$ (1.9)

So, Equation (1.8) can be rewritten as:

$$m\dot{x}_2(t) + cx_2(t) + kx_1(t) = f(t)$$ (1.10)

or

$$\dot{x}_2(t) = -\frac{c}{m}x_2(t) - \frac{k}{m}x_1(t) + \frac{1}{m}f(t).$$ (1.11)

So, Equation (1.8) can be written as a coupled system of two first-order differential equations, that is,

$$\dot{x}_1(t) = x_2(t)$$
$$\dot{x}_2(t) = -\frac{c}{m}x_2(t) - \frac{k}{m}x_1(t) + \frac{1}{m}f(t).$$ (1.12)

The output is the mass displacement

$$y(t) = x_1(t).$$ (1.13)

It is quite common to show the input vector with $u(t)$ instead of $f(t)$. Using this convention, Equation (1.12) can be rewritten as:

$$\dot{x}_1(t) = x_2(t)$$
$$\dot{x}_2(t) = -\frac{c}{m}x_2(t) - \frac{k}{m}x_1(t) + \frac{1}{m}u(t).$$ (1.14)

Finally, the state-space model of system shown in Fig. 1.2 is

$$
\begin{bmatrix} \dot{x}_1(t) \\ \dot{x}_2(t) \end{bmatrix} = \begin{bmatrix} 0 & 1 \\ -\dfrac{k}{m} & -\dfrac{c}{m} \end{bmatrix} \begin{bmatrix} x_1(t) \\ x_2(t) \end{bmatrix} + \begin{bmatrix} 0 \\ \dfrac{1}{m} \end{bmatrix} u(t)
$$

$$
y(t) = [1 \ 0] \begin{bmatrix} x_1(t) \\ x_2(t) \end{bmatrix} + [0]u(t).
$$

(1.15)

According to (1.15), $m = p = 1$, $n = 2$, and

$$
A = \begin{bmatrix} 0 & 1 \\ -\dfrac{k}{m} & -\dfrac{c}{m} \end{bmatrix} \quad B = \begin{bmatrix} 0 \\ \dfrac{1}{m} \end{bmatrix} \quad C = [1 \ 0] \quad D = 0.
$$

(1.16)

1.4   ENTERING THE STATE-SPACE MODEL INTO MATLAB

In the MATLAB command prompt, type the `edit` command—see Fig. 1.4.

After pressing the Enter key, the MATLAB opens an editor for you (see Fig. 1.5). Type the commands shown in Fig. 1.6. Press Ctrl+s in order to save the entered commands in a file. If after pressing Ctrl+s, the message box shown in Fig. 1.7 appeared, click the "Add to Path" and save the file with your desired name. Here we saved the file as "example1.m" (Fig. 1.8). The MATLAB files are saved with the .m extension.

Press F5 in order to run the code. After running the code, new variables are added to the workspace (Fig. 1.9). The variable sys, is of 1×1 ss type. This shows that variable sys is a state-space model with 1 input and 1 output. A system with 1 input and 2 output is shown as 2×1 ss in the MATLAB Workspace.

In the Command Window, type sys and press the Enter key. As seen in Fig. 1.10, the MATLAB shows the value of matrices A, B, C, and D of the state-space model. For instance, if you need the value of matrix A, you can simply type `sys.A` or `sys.a` in MATLAB command window. You can extract the matrices A, B, C, and D of state-space model with the aid of `ssdata` command, as shown in Fig. 1.12.

If you need to see the number of outputs, inputs, and states, type size(sys) in the Command Window (see Fig. 1.13). You can obtain the `order` of a system with the aid of order command (Fig. 1.14).

The order of a dynamic system model is the number of poles (for proper transfer functions) or the number of states (for state-space models). For improper transfer functions, the order is defined as the minimum number of states needed to build an equivalent state-space model (ignoring pole/zero cancellations).

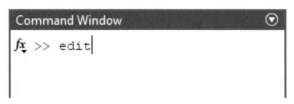

Figure 1.4: Running the editor.

Figure 1.5: The editor window MATLAB.

You can obtain a minimal state-space representation for a given transfer function using the ss command as well. Assume that $H(s)$ is given as:

$$H(s) = \begin{bmatrix} \dfrac{s+1}{s^3 + 3s^2 + 3s + 2} \\ \dfrac{s^2 + 3}{s^2 + s + 1} \end{bmatrix}. \tag{1.17}$$

Take a look to the commands shown in Fig. 1.15. Both sys2 and sys3 are the state-space representation of $H(s)$, i.e., $(C(sI - A)^{-1}B + D = H(s))$. However, sys2 has 5 states, while sys3 has only 3 states. sys2 is not the minimal realization of $H(s)$. Using the ss command with the "minimal" part, produces a minimal realization.

You can obtain the transfer function associated this state-space models using the tf command (Fig. 1.16).

```
Editor - Untitled                                        ⊙ ✕
   Untitled  ✕   +
1 —     k=1; %defining the value of k
2 —     m=2; %defining the value of m
3 —     c=3; %defining the value of c
4
5 —     A=[0 1;-k/m -c/m];
6 —     B=[0;1/m];
7 —     C=[1 0]
8 —     D=[0]
9
10 —    sys=ss(A,B,C,D);
11
```

Figure 1.6: Defining the state-space model of system shown in Fig. 1.2.

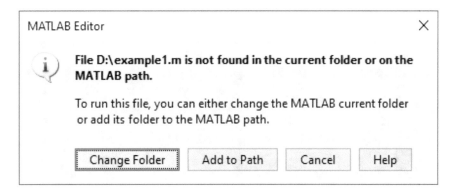

Figure 1.7: The warning message. Click Add to Path in order to run your program.

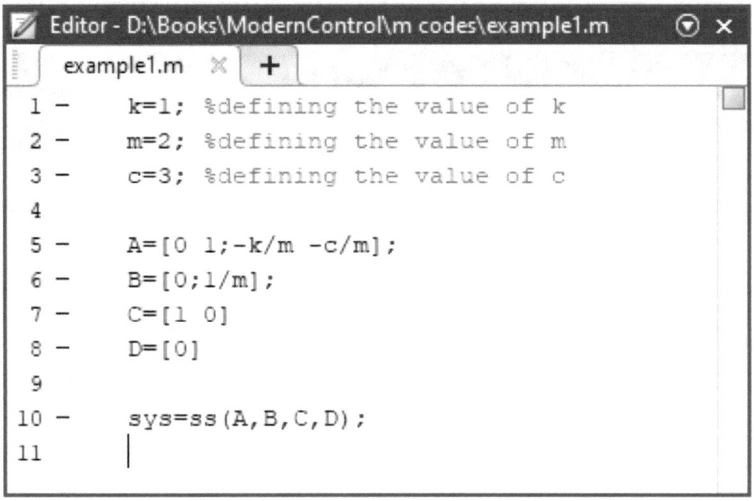

```
Editor - D:\Books\ModernControl\m codes\example1.m        ⊙ ✕
 example1.m   ✕   +
 1 -      k=1; %defining the value of k
 2 -      m=2; %defining the value of m
 3 -      c=3; %defining the value of c
 4
 5 -      A=[0 1;-k/m -c/m];
 6 -      B=[0;1/m];
 7 -      C=[1 0]
 8 -      D=[0]
 9
10 -      sys=ss(A,B,C,D);
11        |
```

Figure 1.8: The code is saved with the name example1.m.

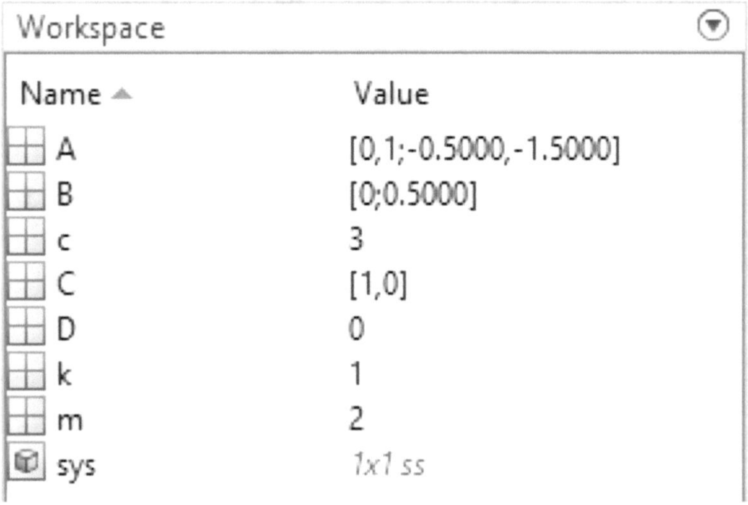

Workspace   ⊙

Name ▲	Value
A	[0,1;-0.5000,-1.5000]
B	[0;0.5000]
c	3
C	[1,0]
D	0
k	1
m	2
sys	1x1 ss

Figure 1.9: Variables in the workspace of MATLAB.

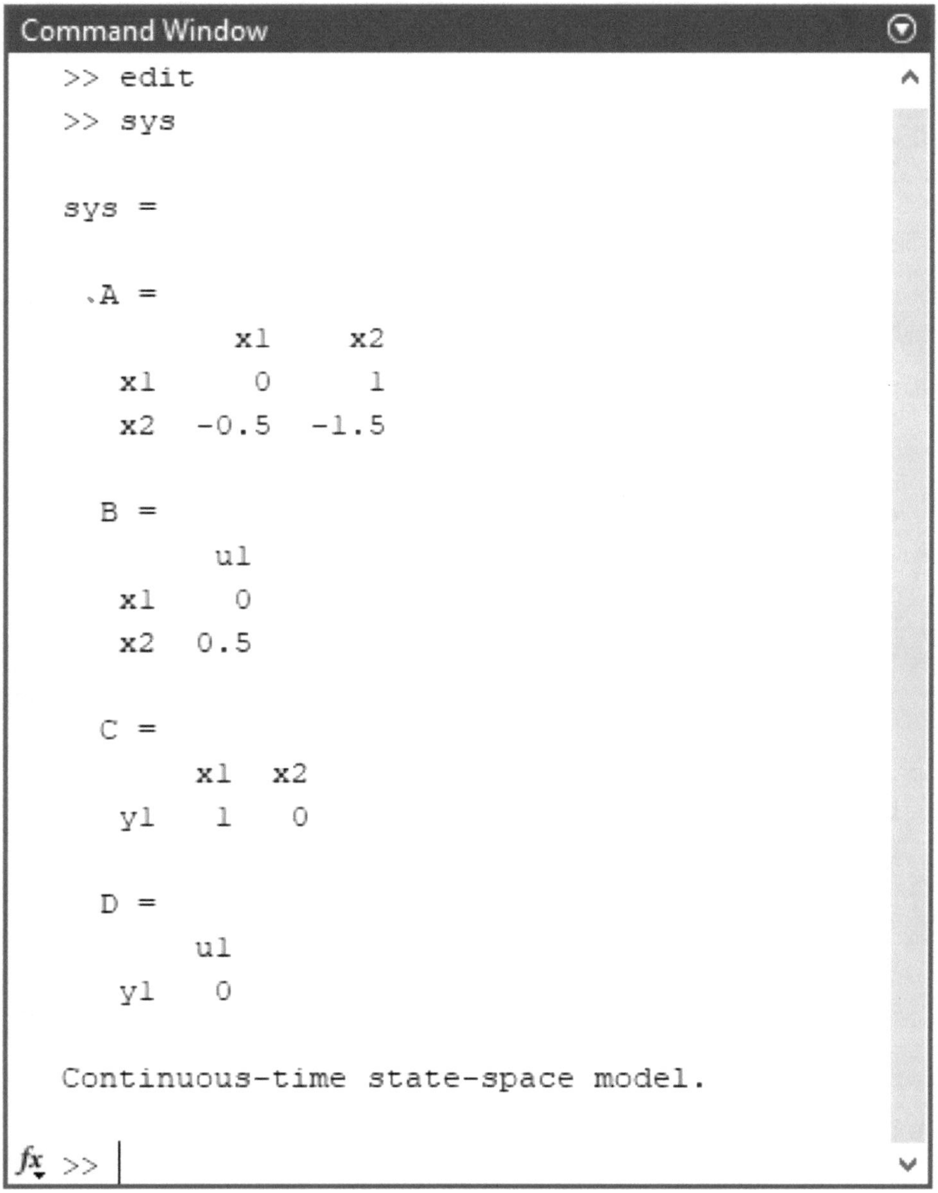

```
Command Window

    >> edit
    >> sys

  sys =

   .A =
               x1       x2
      x1        0        1
      x2      -0.5     -1.5

    B =
               u1
      x1        0
      x2       0.5

    C =
              x1    x2
      y1       1     0

    D =
              u1
      y1       0

  Continuous-time state-space model.

fx  >> |
```

Figure 1.10: Value of matrices A, B, C, and D for state-space model sys.

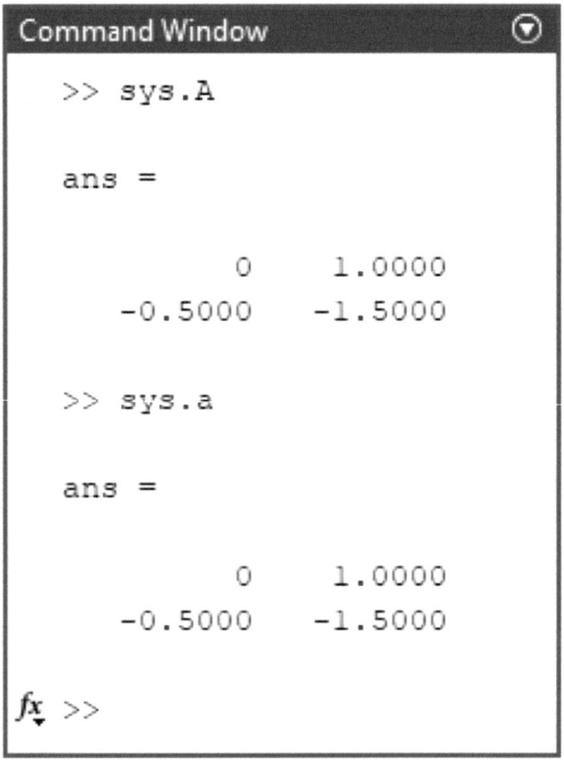

Figure 1.11: Obtaining the value of matrix A.

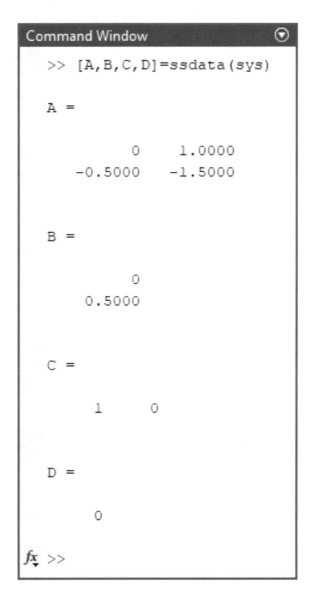

Figure 1.12: Obtaining the value of matrices A, B, C, and D with the aid of ssdata command.

```
Command Window                                              ⊙
    >> size(sys)
    State-space model with 1 outputs, 1 inputs, and 2 states.
fx >>
```

Figure 1.13: Obtaining the number outputs, inputs, and states with the aid of size command.

Figure 1.14: Obtaining the order of a system with the aid of system command.

```
Command Window                                              ⊙
    >> H = [tf([1 1],[1 3 3 2]) ; tf([1 0 3],[1 1 1])];
    >> sys2=ss(H);
    >> size(sys2)
    State-space model with 2 outputs, 1 inputs, and 5 states.
    >> sys3=ss(H,'minimal');
    >> size(sys3)
    State-space model with 2 outputs, 1 inputs, and 3 states.
fx >> |
```

Figure 1.15: Obtaining the minimal state-space realization of a system.

```
Command Window                                    ⊙

  >> tf(sys2)

  ans =

    From input to output...
                  s + 1
      1:  ---------------------
          s^3 + 3 s^2 + 3 s + 2


            s^2 + 3
      2:  -----------
          s^2 + s + 1

  Continuous-time transfer function.

  >> tf(sys3)

  ans =

    From input to output...
                  s + 1
      1:  ---------------------
          s^3 + 3 s^2 + 3 s + 2


          s^3 + 2 s^2 + 3 s + 6
      2:  ---------------------
          s^3 + 3 s^2 + 3 s + 2

  Continuous-time transfer function.

fx  >>
```

Figure 1.16: The transfer function of both systems are the same.

```
Command Window                              ⊙

    >> pole(sys)

    ans =

        -0.5000
        -1.0000

    >> zero(sys)

    ans =

      0×1 empty double column vector

fx >> |
```

Figure 1.17: Obtaining the poles and zeros of a system.

Figure 1.18: The step response of a system can be drawn with the aid of step command.

You can obtain the poles and zeros of defined system with the aid of pole and zero commands. This system has no zero since the denominator has no roots (Fig. 1.17).

You can see the system step response using the step command (Fig. 1.18). The step response of system is shown in Fig. 1.19. As seen in Fig. 1.19, there is no overshoot in the response. This is expected since the system poles are not complex.

MATLAB has a powerful command named lsim for simulating the linear systems with arbitrary inputs. The lsim command takes the time vector and value of system input for each time instant. The commands shown in Fig. 1.20 stimulate the system sys with an array of 1's. The variable t is the time vector and u is the vector of input values for each time instant. This input acts like a step input. That is why result of simulation (Fig. 1.21) is the same as with

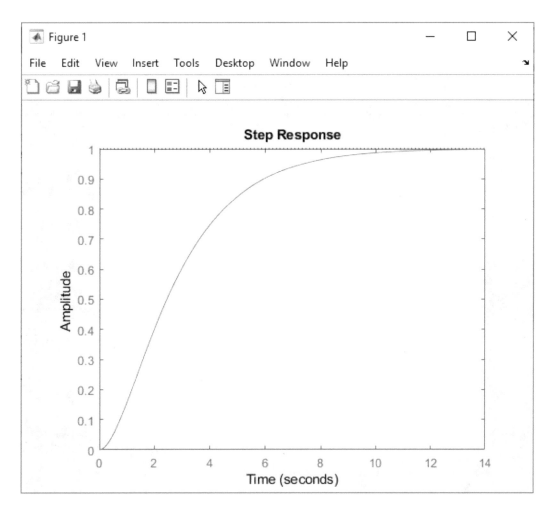

Figure 1.19: The step response of system sys.

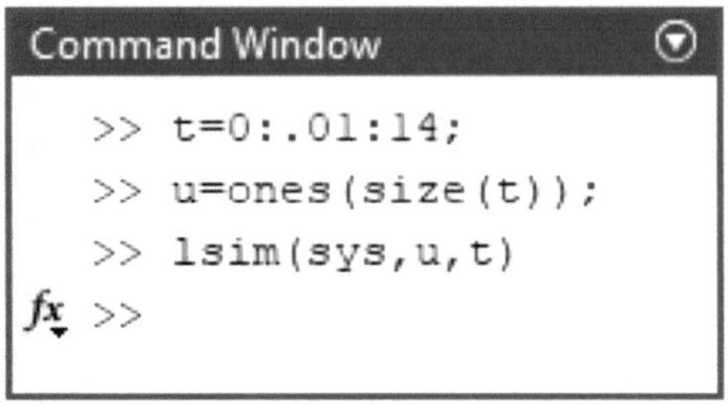

Command Window

```
>> t=0:.01:14;
>> u=ones(size(t));
>> lsim(sys,u,t)
fx >>
```

Figure 1.20: Using the lsim command to draw the step response.

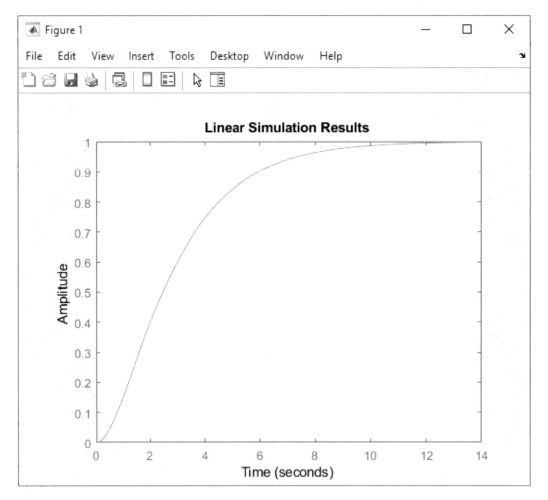

Figure 1.21: Result of running the command shown in Fig. 1.20.

```
Command Window                          ⊙
    >> t=0:.01:14;
    >> u=sin(t);
    >> lsim(sys,u,t,[-1;2])
 fx >> |
```

Figure 1.22: Simulating the system with a sinusoidal input and an initial condition.

Fig. 1.19. If you have a non-zero initial conditions, you can enter it to the lsim command as well. For instance, the command shown in Fig. 1.22, simulate the system with a sinusoidal input and initial conditions $x_1 = -1$ and $x_2 = 2$. Result of this command is shown in Fig. 1.23.

If you want, you can generate a random state-space model in MATLAB. In order to do this, use the rss command. For instance, in order to produce a state-space model, with 4 states, 2 outputs, and 3 inputs, write rss(4,2,3). Sample output for the rss(4,2,3) command is shown in Fig. 1.24. Since the rss command produces a random system, result of running rss(4,2,3) command on your computer may not be the same as the one shown in Fig. 1.24.

1.5   ENTERING THE STATE-SPACE MODELS TO SIMULINK

The state-space models can be entered to Simulink as well. In order to do this, type simulink in the MATLAB Command Window and press the Enter key. In the opened window (Fig. 1.25), click the "Blank Model."

In the opened window, click the "Library Browser" (Fig. 1.26). After Library Browser window appeared, click the Continuous category (Fig. 1.27). Drag and drop the state-space block (Fig. 1.28) to the model. Double click the drag and dropped state-space model and enter the matrices A, B, C, and D in the boxes. If the system starts from a nonzero state, i.e.,

$$\dot{x} = Ax + Bu$$
$$y = Cx + Du \qquad (1.18)$$
$$x_{t=t_0} = x_0,$$

then enter this value in the Initial conditions box.

In order to see the step response of system, connect a step block to input of system. Step block can be found in the Sources category of Simulink Library Browser window.

Double click the step block and change the Step time to 0 (Fig. 1.32), so the jump from initial value (0) to final value (1) happens at $t = 0$.

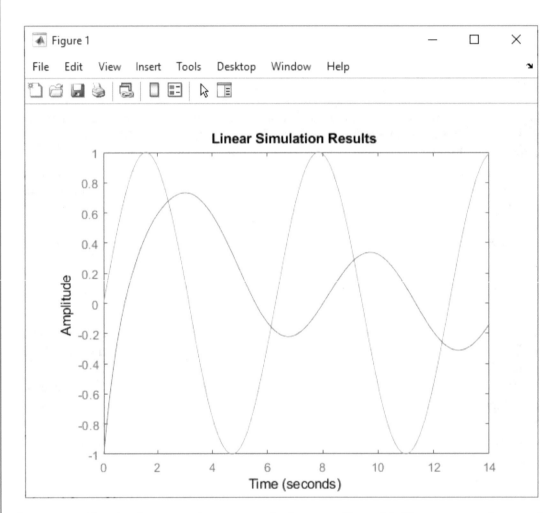

Figure 1.23: Result of running the commands shown in Fig. 1.22. The grey waveform is the system input. The blue one is the system response.

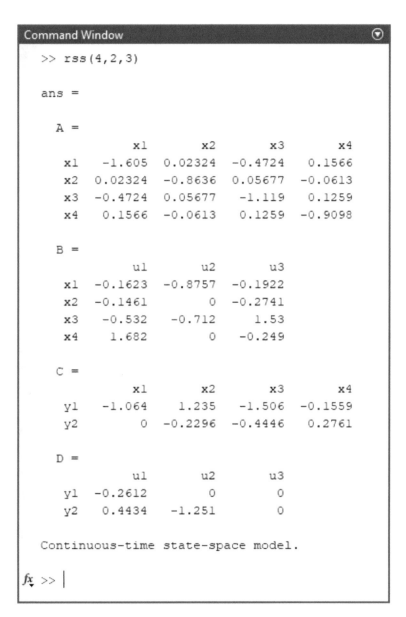

```
Command Window                                        ⊙

  >> rss(4,2,3)

  ans =

   A =
              x1        x2        x3        x4
       x1   -1.605   0.02324   -0.4724    0.1566
       x2   0.02324  -0.8636   0.05677   -0.0613
       x3   -0.4724  0.05677    -1.119    0.1259
       x4    0.1566  -0.0613    0.1259   -0.9098

   B =
              u1        u2        u3
       x1   -0.1623  -0.8757   -0.1922
       x2   -0.1461        0   -0.2741
       x3    -0.532   -0.712      1.53
       x4     1.682        0    -0.249

   C =
              x1        x2        x3        x4
       y1   -1.064     1.235    -1.506   -0.1559
       y2        0   -0.2296   -0.4446    0.2761

   D =
              u1        u2        u3
       y1   -0.2612        0         0
       y2    0.4434    -1.251        0

  Continuous-time state-space model.

fx >> |
```

Figure 1.24: Sample output for rss(4,2,3).

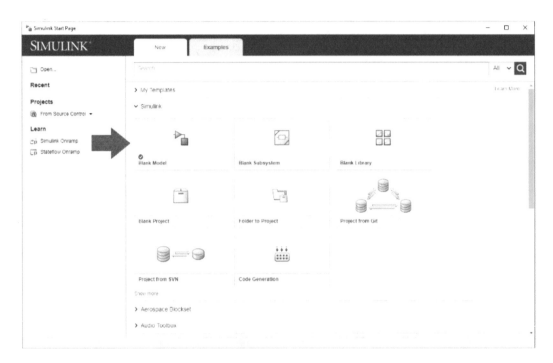

Figure 1.25: Simulink start page.

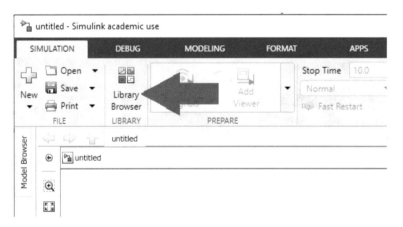

Figure 1.26: Place of the library browser button.

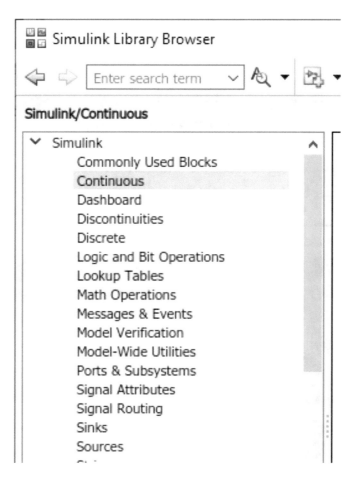

Figure 1.27: The continuous section of Library Browser window.

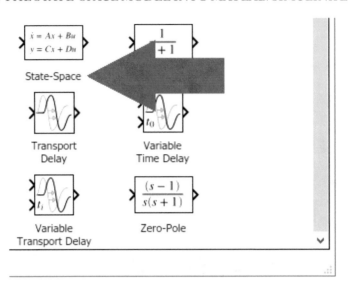

Figure 1.28: State-space block.

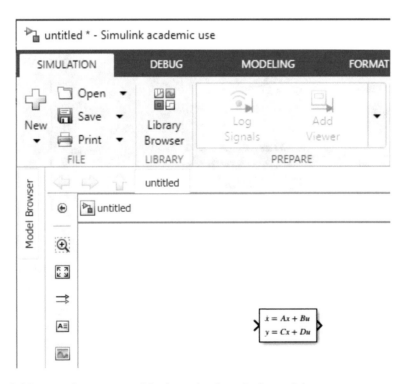

Figure 1.29: Addition of state-space block to the Simulink model.

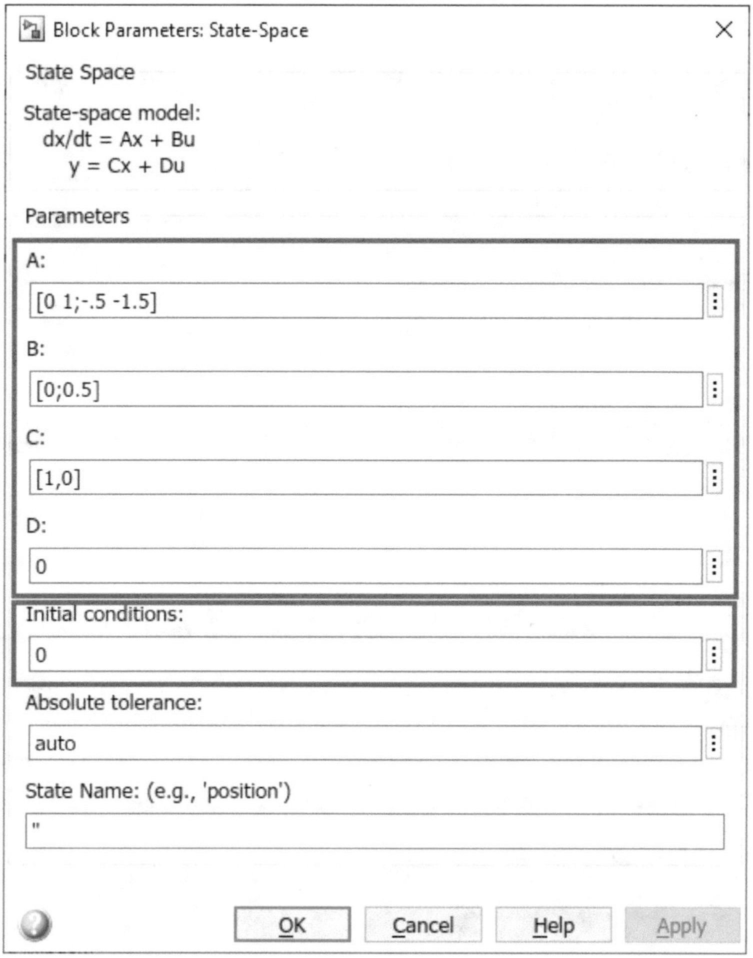

Block Parameters: State-Space   ✕

State Space

State-space model:
    dx/dt = Ax + Bu
        y = Cx + Du

Parameters

A:
[0 1;-.5 -1.5]

B:
[0;0.5]

C:
[1,0]

D:
0

Initial conditions:
0

Absolute tolerance:
auto

State Name: (e.g., 'position')
"

OK   Cancel   Help   Apply

Figure 1.30: Settings of state-space blocks.

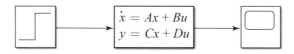

Figure 1.31: Connecting the step block to input of stat-space block.

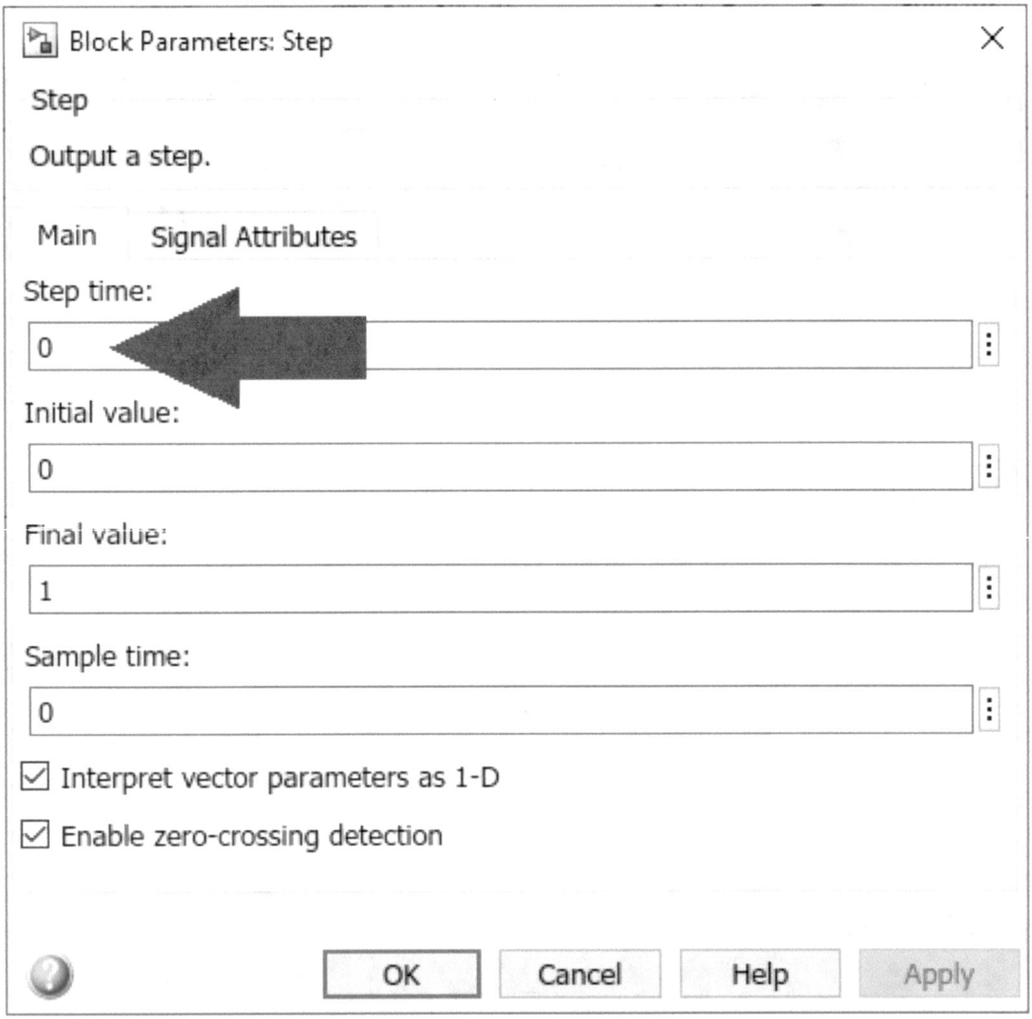

Figure 1.32: **Settings of step block.**

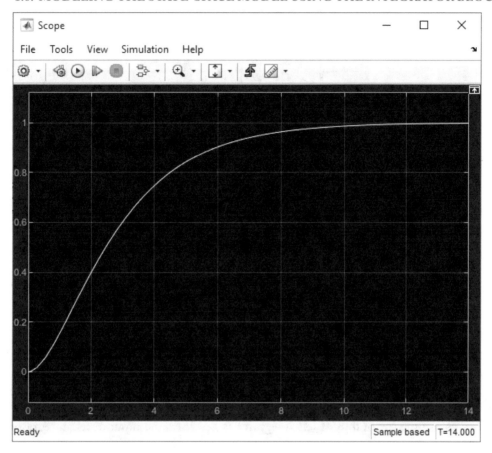

Figure 1.33: Simulation results.

Simulation result is shown in Fig. 1.33. This result is the same as the one in Fig. 1.19. If your system has more than one input and more than one output, then you can use the multiplexer and demultiplexer blocks for entering the signal into the system and getting out the signal from the system (Fig. 1.34).

1.6   MODELING THE STATE-SPACE MODEL USING THE INTEGRATOR BLOCK

In the previous section, we used the state-space block in the Simulink model. If you want to realize a nonlinear state-space model, then you cannot use the state-space block. You need to use the integrator block and you need to draw the system diagram from scratch. This system can be used with linear systems as well. As our first example, let us draw the block diagram of me-

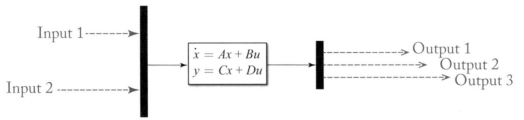

Figure 1.34: Entering multiple signals to input/output of system with the aid of multiplexer/de-multiplexer blocks.

chanical system shown in Fig. 1.2. The state-space model of system is given in Equation (1.14). The state equation is rewritten below:

$$\begin{cases} \dot{x}_1(t) = x_2(t) \\ \dot{x}_2(t) = -\dfrac{c}{m}x_2(t) - \dfrac{k}{m}x_1(t) + \dfrac{1}{m}u(t). \end{cases} \tag{1.14}$$

The state equation has two state variables. So we need two integrators in the model. The integrator block can be found in the Continuous section of Simulink Library Browser (Fig. 1.35).

Drag and drop two integrators to the Simulink model (Fig. 1.36). Let us assume the output of right integrator is $y = x_1$. Then the input of right integrator is $\dot{y} = \dot{x}_1$ because the input is the derivative of output (Fig. 1.37).

Let us assume the output of left integrator is x_2. Then the input of left integrator is $\dot{x}_2$ (Fig. 1.38).

According to the state equation, $\dot{x}_1 = x_2$, so we can connect the input of right integrator ($\dot{x}_1$) to the output of right integrator (x_2) (Fig. 1.39).

Second state equation ($\dot{x}_2(t) = -\frac{c}{m}x_2(t) - \frac{k}{m}x_1(t) + \frac{1}{m}u(t)$) tell us that, $\dot{x}_2$ (input of first integrator) is the summation of x_2 (output of left integrator) with coefficient $-\frac{c}{m}$, x_1 (output of right integrator) with coefficient $-\frac{k}{m}$ and external input of the system with coefficient $\frac{1}{m}$. This is done with the aid blocks shown in Fig. 1.40.

In order to see the output of system, we need to add a scope block to the output of system (in this example, output of right integrator) (Fig. 1.41).

As another example, we draw the Simulink diagram of following equation (Van der Pol equation):

$$\ddot{x} - m\left(1 - x^2\right)x + \dot{x} = 0. \tag{1.19}$$

This equation is a non-conservative oscillator with nonlinear damping. m is a scalar parameter indicating the nonlinearity and the strength of damping. When $m = 0$, the given system is linear. This equation has no external input.

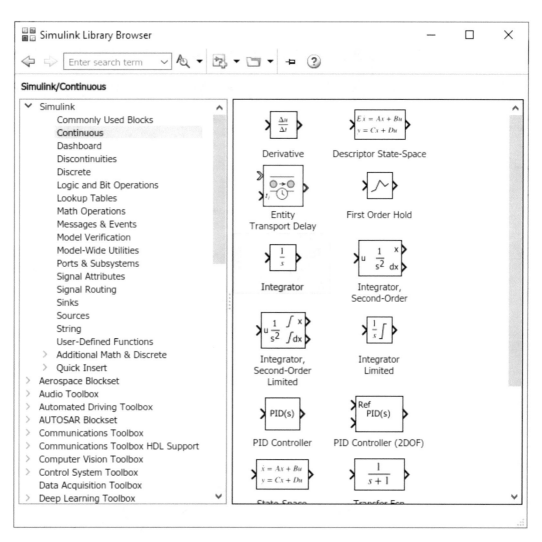

Figure 1.35: The integrator block is placed in the Continuous section of Simulink library browser.

Figure 1.36: Addition of two integrators to the model.

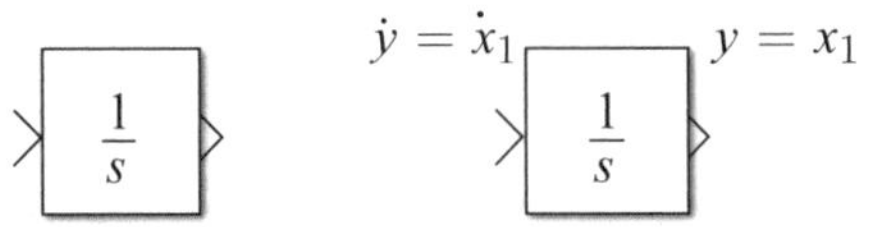

$$\dot{y} = \dot{x}_1 \qquad y = x_1$$

Figure 1.37: Defining the variable x_1 and $\dot{x}_1$.

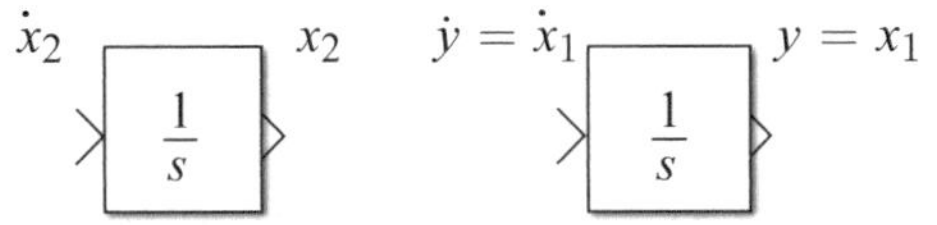

$$\dot{x}_2 \qquad x_2 \qquad \dot{y} = \dot{x}_1 \qquad y = x_1$$

Figure 1.38: Defining the variable x_2 and $\dot{x}_2$.

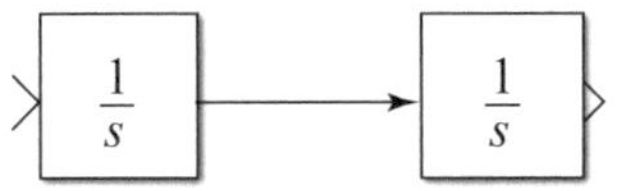

Figure 1.39: Realization of equation $\dot{x}_1 = x_2$.

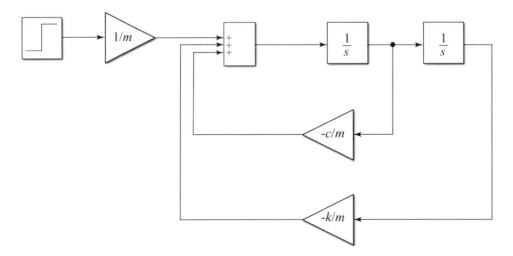

Figure 1.40: Realization of equation $\dot{x}_2(t) = -\frac{c}{m}x_2(t) - \frac{k}{m}x_1(t) + \frac{1}{m}u(t)$.

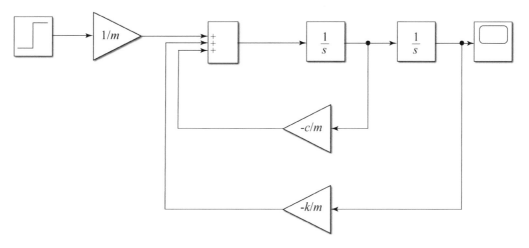

Figure 1.41: Addition of scope block to the system.

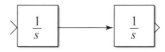

Figure 1.42: Defining the variable x_1, x_2, $\dot{x}_1$, and $\dot{x}_2$.

Figure 1.43: Realization of equation $\dot{x}_1 = x_2$.

By choosing the $x_1 = x$ and $x_2 = \dot{x}$ we obtain:

$$\begin{cases} \dot{x}_1 = x_2 \\ \dot{x}_2 = m \left(1 - x_1^2\right) x_2 - x_1, \end{cases} \tag{1.20}$$

and the output equation is $y = x = x_1$. Since we have two state equations, we need two integrators (Fig. 1.42).

The first state equation ($\dot{x}_1 = x_2$) is satisfied by connecting the output of left integrator (x_2) to input of right integrator ($\dot{x}_1$) (Fig. 1.43). The second state equation ($\dot{x}_2 = m \left(1 - x_1^2\right) x_2 - x_1$) require multiplication and square of signal x_1. The multiplication block can be found in the Math Operation section of Simulink Library Browser (Fig. 1.44). In order to produce the term x_1^2, the signal x_1 is multiplied by itself.

The completed Simulink diagram of given system is shown in Fig. 1.45.

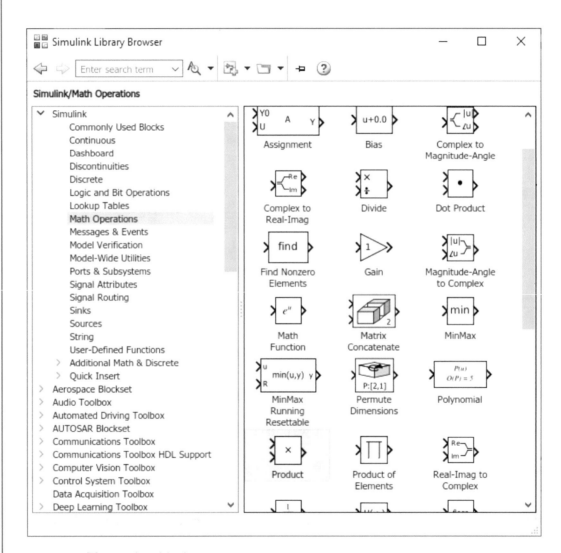

Figure 1.44: The product block.

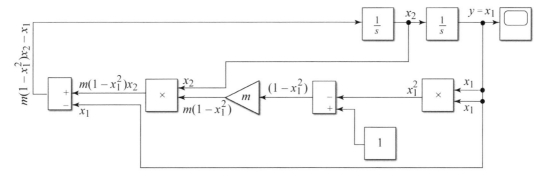

Figure 1.45: Simulink diagram of state-space model given in Equation (1.15).

Let us simulate the system for $m = 1$ and initial conditions $x_{1,0} = 1$ and $x_{2,0} = 1$. In order to enter the initial condition $x_{1,0} = 1$, double click the right integrator and write 1 in the Initial condition box (Fig. 1.46).

In order to enter the initial condition $x_{2,0} = 2$, double click the right integrator and write 2 in the Initial condition box (Fig. 1.47). You can define the value of variable m in the MATLAB Workspace (Fig. 1.48). Simulation result is shown in Fig. 1.49.

We can draw the phase portrait (see Chapter 2 of [1] or Chapter 1 of [2]) of the system using the XY Graph block. The XY Graph block can be found in the Sink section of Simulink Library Browser (Fig. 1.50).

The XY Graph block has two inputs. The upper input controls the X axis and the lower input controls the Y axis. The XY Graph acts like the XY mode of oscilloscopes (Fig. 1.51). Connect the upper input to x_1 (output of right integrator) and lower input to x_2 (output of left integrator).

Change the boxes as shown in Fig. 1.52 in order to see the drawn curve completely. The simulation result is shown in Fig. 1.53. The phase portrait shown in Fig. 1.53 starts from initial state (point (1,2)) and does not converge to any specific state as time increases. This is expected from the time response of system (Fig. 1.49), because it is oscillatory. You can change the value of m and see its effect on the system responses.

1.7   SIMULATION OF STATE-SPACE MODEL WITH THE AID OF MATLAB FUNCTION BLOCK

In the previous section, we draw the Simulink diagram with the aid of blocks available in Simulink. Using this method (specially for high-order systems with many states) may result into a crowded model which is difficult to understand and modify. The MATLAB function block (Fig. 1.54) provides an easy solution for this problem.

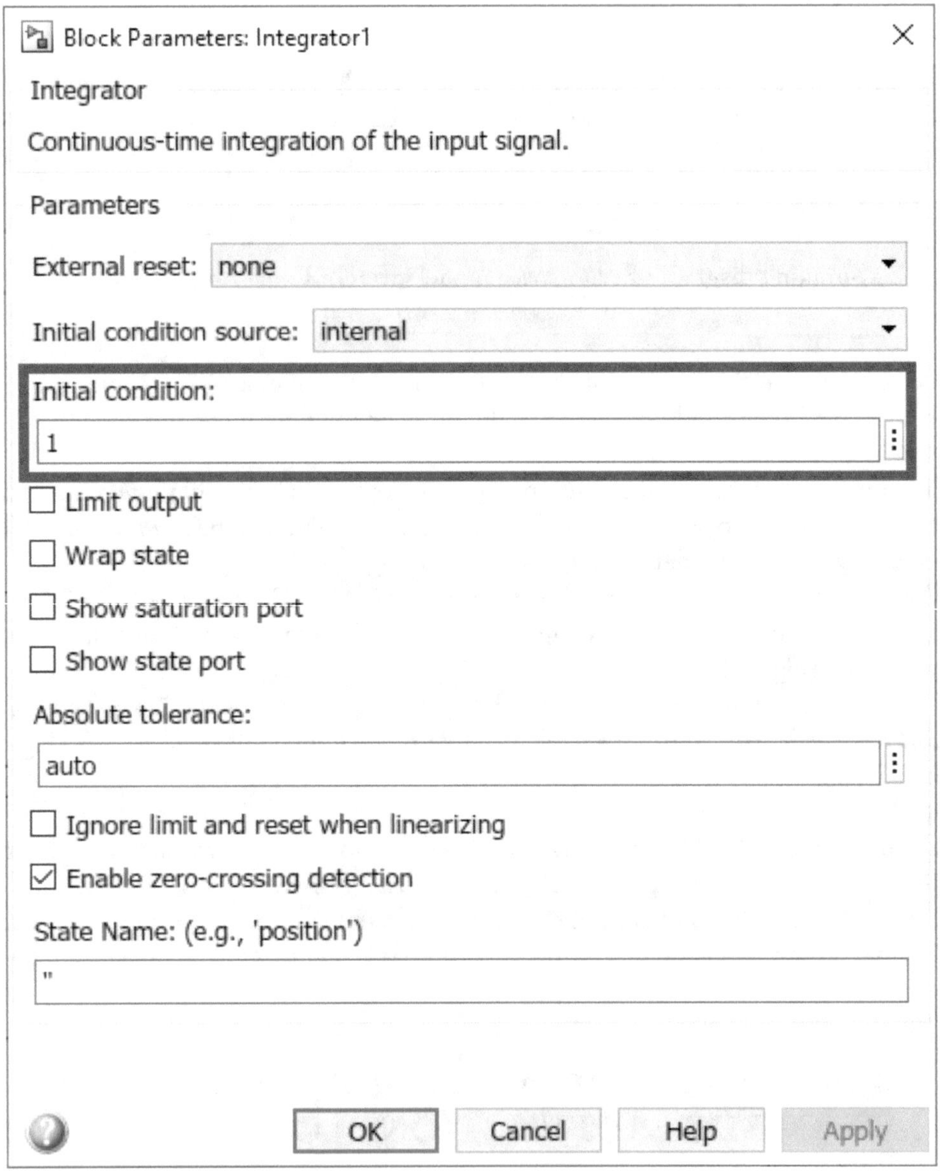

Block Parameters: Integrator1   ✕

Integrator

Continuous-time integration of the input signal.

Parameters

External reset: none   ▼

Initial condition source: internal   ▼

Initial condition:

1   ⋮

☐ Limit output

☐ Wrap state

☐ Show saturation port

☐ Show state port

Absolute tolerance:

auto   ⋮

☐ Ignore limit and reset when linearizing

☑ Enable zero-crossing detection

State Name: (e.g., 'position')

"

          OK   Cancel   Help   Apply

Figure 1.46: Setting the initial condition of right integrator.

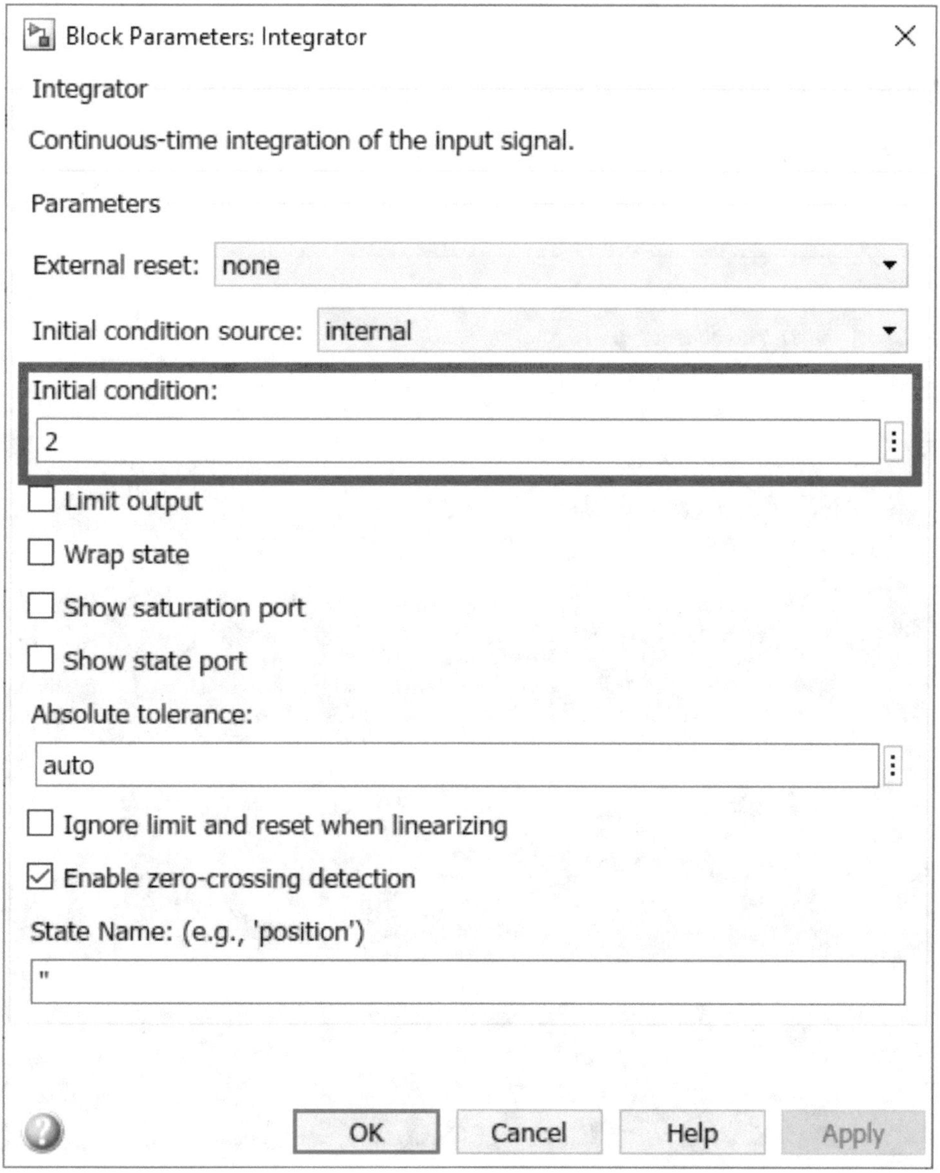

Figure 1.47: Setting the initial condition of left integrator.

Command Window

fx >> m=1;

Figure 1.48: Defining the value of *m* in the workspace of MATLAB.

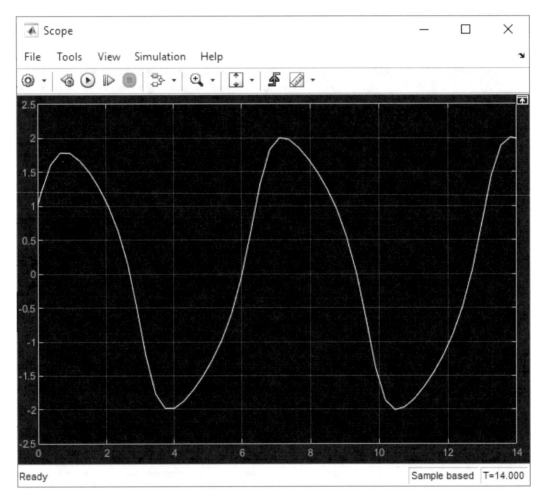

Figure 1.49: Simulation result.

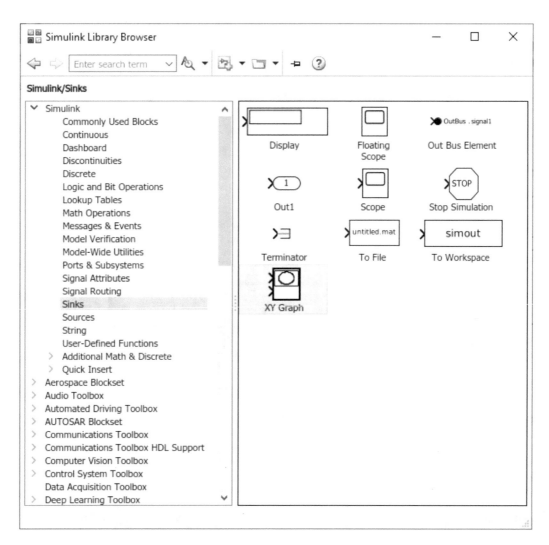

Figure 1.50: The XY graph block.

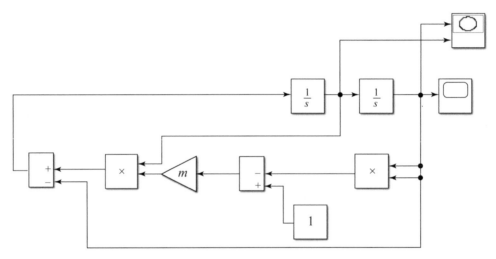

Figure 1.51: Addition of XY graph block to the model.

Figure 1.52: Setting of XY graph block.

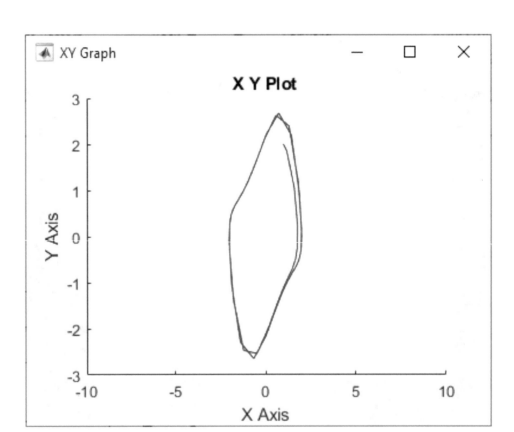

Figure 1.53: The phase portrait of system.

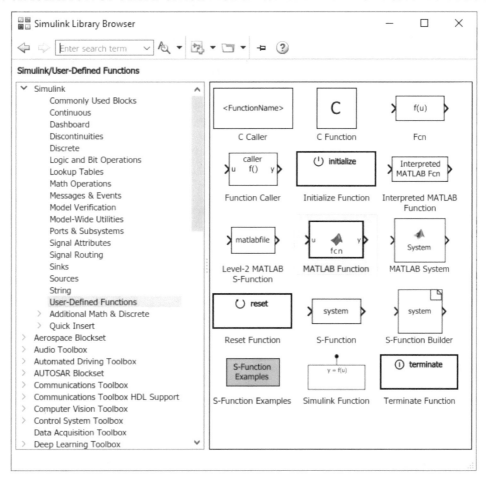

Figure 1.54: The MATLAB function block.

As an example assume that we want to simulate the Van der Pol equation with initial condition (1,2). In order to do this, we add an integrator block, a MATLAB function block and a scope block to the model. We do the connections as shown in Fig. 1.55.

Double click the integrator block and write [1;2] in the Initial condition box. This tells the MATLAB that this integrator has two outputs. Using this technique, we can do the job of two integrators with only one integrator symbol on the model. In order to understand this matter consider the simple simulation diagram shown in Fig. 1.57. Here, two constant inputs (with values of 1 and 2) to the integrator by means of a multiplexer block. After running the simulation, the result shown in Fig. 1.58 appears. This is expected since the $\int 1 dt = t + t_{0,1} = t + 1$ (yellow signal) and $\int 2 dt = 2t + t_{0,2} = 2t + 2$ (blue signal). $t_{0,1}$ and $t_{0,2}$ are the initial

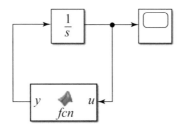

Figure 1.55: Connecting the MATLAB function block to the integrator.

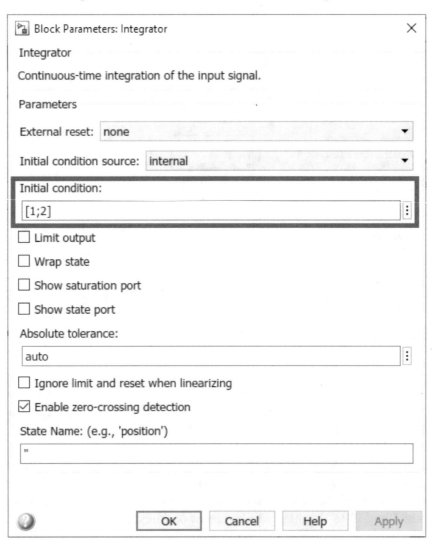

Figure 1.56: Defining the initial condition for the integrator block.

conditions defined by the user. Note that the integrator block in Fig. 1.57 has the same settings with the one shown in Fig. 1.56.

Double click the MATLAB function block in Fig. 1.55 and write the code shown in Fig. 1.59. This code is the MATLAB translation of Equation (1.19). u(1) shows the first input to MATLAB function block. In the same way, u(2) shows the second input to MATLAB function block.

The input for MATLAB function block is the output of integrator. So, u(1), which shows the first input of MATLAB function block, is the first output of integrator. In the same way, u(2), equals the second output of integrator.

After running the simulation, the result shown in Fig. 1.60 is obtained. You can obtain the phase portrait of system using the diagram shown in Fig. 1.61.

After running the simulation, the result shown in Fig. 1.62 is obtained. This result is the same as the one shown in Fig. 1.53.

You can add one more scope if you want to see only the first state variable, i.e., x_1 (Fig. 1.63). The waveform of this new scope is shown in Fig. 1.64. This result is the same as the one obtained before (see Fig. 1.49).

The previous example had no external input. Let's study a system with external input. Assume that we want to simulate the following system:

$$\begin{cases} \dot{x}_1 = x_2 \\ \dot{x}_2 = \left(1 - x_1^2\right) x_2 - x_1 + 2u. \end{cases} \tag{1.21}$$

The input u is a step function. The initial conditions are zero. The external input is injected to the MATLAB function block using a multiplexer (Fig. 1.65). The code inside the MATLAB Function block is shown in Fig. 1.66. The u(3) in the MATLAB function block shows the external input (in this example step input). The first two inputs come from the integrator, so the external step input is the third input.

The integrator block and step block settings are shown in Figs. 1.67 and 1.68, respectively. Simulation result is shown in Fig. 1.69. This is an expected result since in steady-state, derivatives (left side of Equation (1.21)) become zero. So,

$$\begin{cases} 0 = x_2 \\ 0 = \left(1 - x_1^2\right) x_2 - x_1 + 2u. \end{cases} \tag{1.22}$$

This leads to $x_2 = 0$ and $-x_1 + 2u = 0$, which means $x_1 = 2u$ in steady state. Since the value of step input is 1, the steady-state value of x_1 must be $2 \times 1 = 2$. And this is what Fig. 1.69 shows: the x_1 converged toward 2 and x_2 converged toward 0.

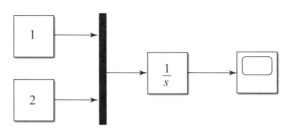

Figure 1.57: Calculating two different integration with one integrator block.

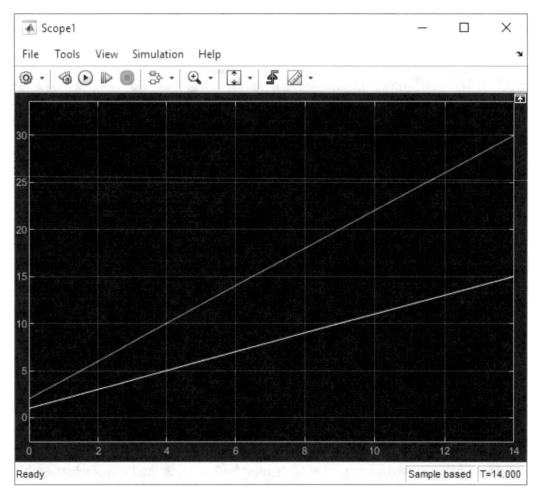

Figure 1.58: Result of simulating the block diagram shown in Fig 1.57.

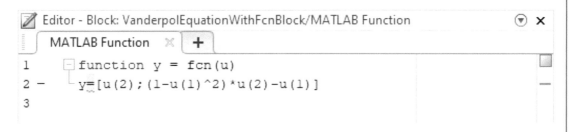

Figure 1.59: Defining the MATLAB function.

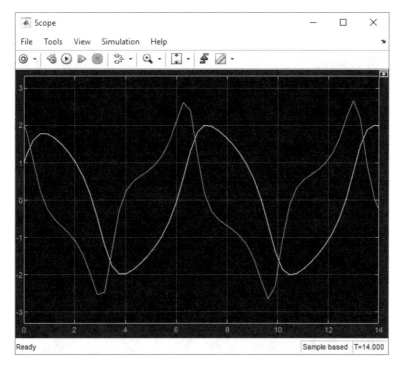

Figure 1.60: Result of simulation.

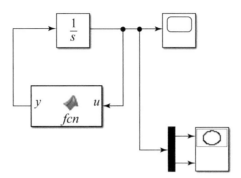

Figure 1.61: Addition of a XY graph block to the model.

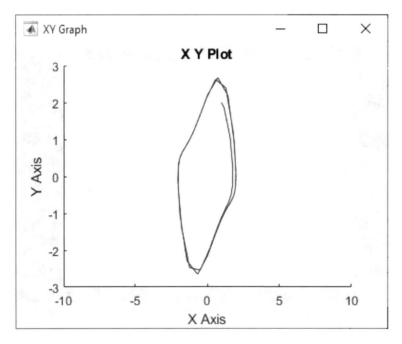

Figure 1.62: The phase portrait of system.

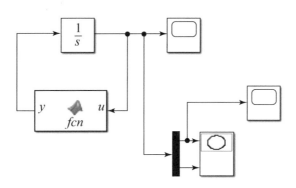

Figure 1.63: A multiplexer could be used to isolate and see the desired signal.

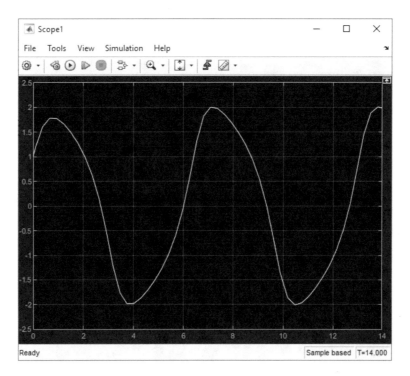

Figure 1.64: Waveform for state x_1.

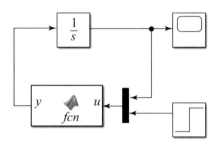

Figure 1.65: Simulation diagram of Equation (1.21).

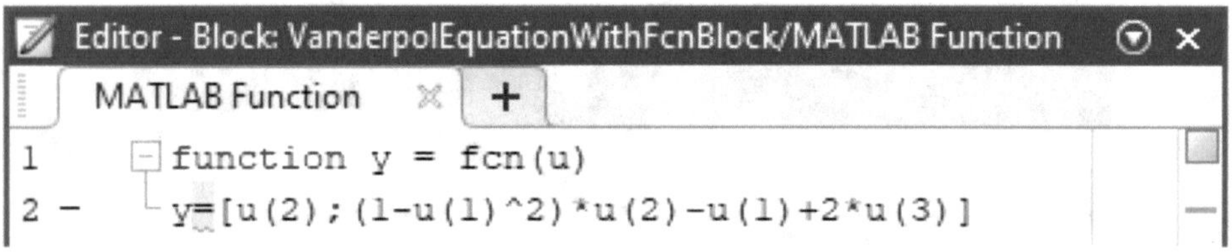

Editor - Block: VanderpolEquationWithFcnBlock/MATLAB Function ⊙ ✕

    MATLAB Function   ✕   +

```
1        function y = fcn(u)
2 -      y=[u(2);(1-u(1)^2)*u(2)-u(1)+2*u(3)]
```

Figure 1.66: Inside of MATLAB function block in Fig. 1.65.

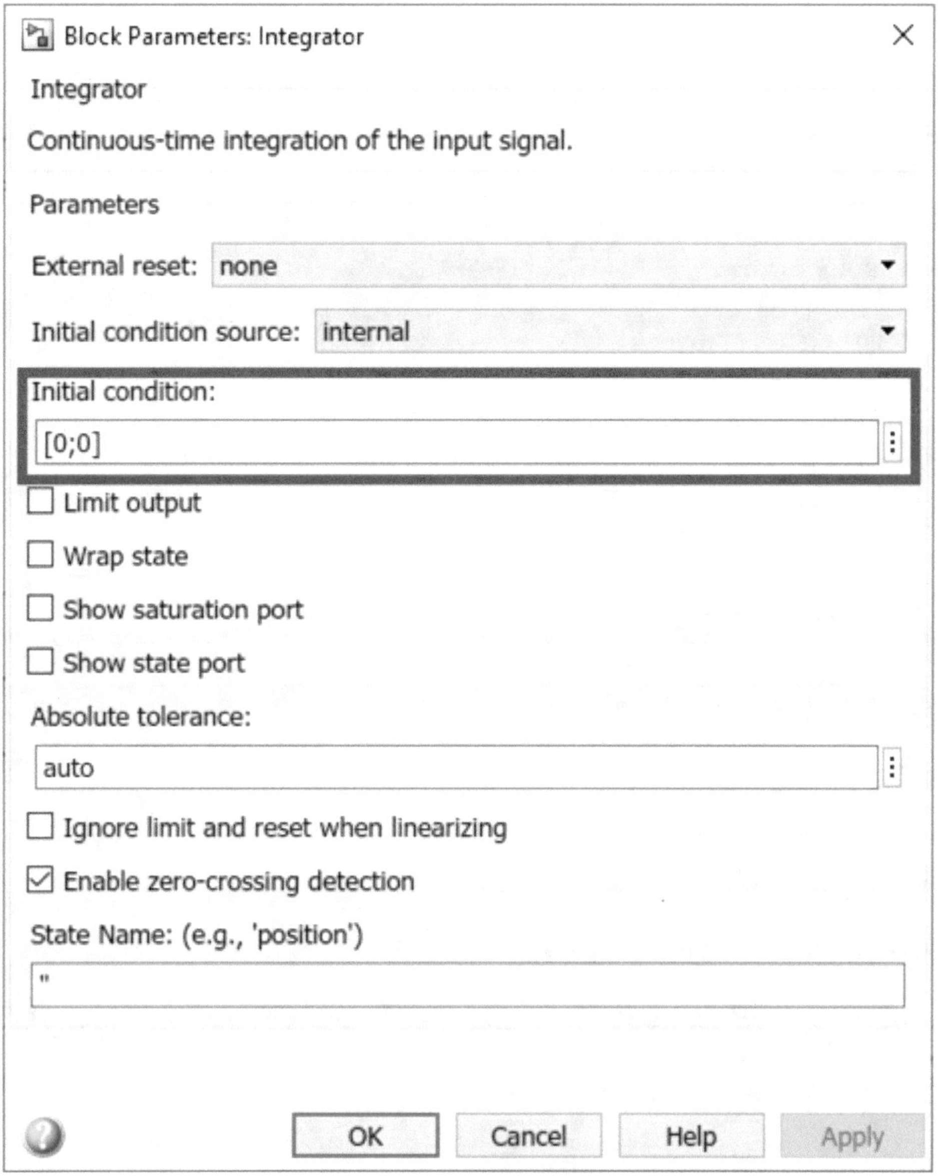

Figure 1.67: Defining the initial condition of integrator block.

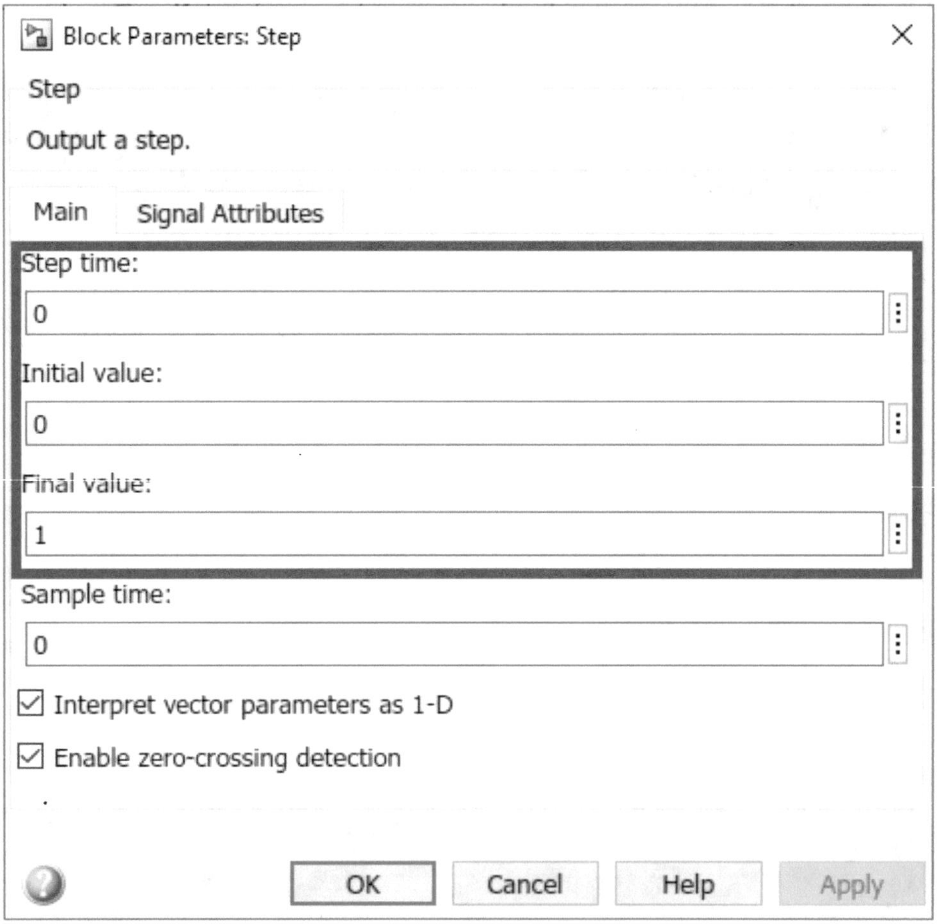

Figure 1.68: Settings of step block.

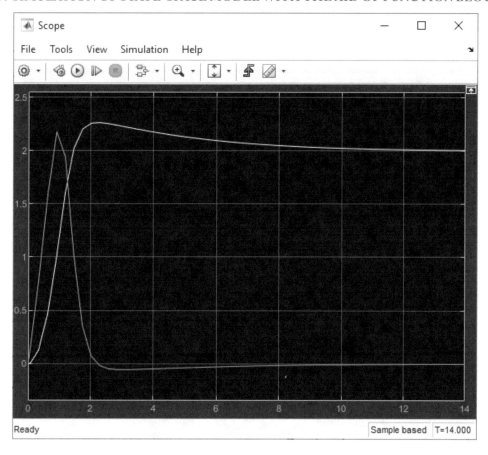

Figure 1.69: Simulation result.

The smoothness of Fig. 1.69 can be increased. In order to increase the smoothness of curves, go to the MODELING section of Simulink and click the Model Settings (Fig. 1.70). After you clicked the Model Settings, a drop-down list will be appeared. Click the Model Settings from the appeared list (Fig. 1.71). In the appeared window, click the Solver (Fig. 1.72).

Enter a small number in the Fixed-step size (fundamental sample time) box. Don't decrease the number too much, because it increase the required time for doing the simulation (Fig. 1.73).

The simulation with Fixed-step size (fundamental sample time) = 0.001 is shown in Fig. 1.74. This figure is smoother in comparison with the one shown in Fig. 1.69.

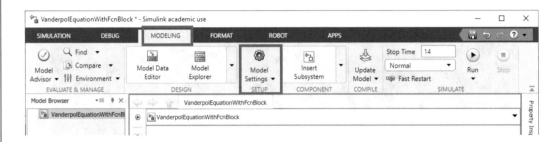

Figure 1.70: **Model settings button.**

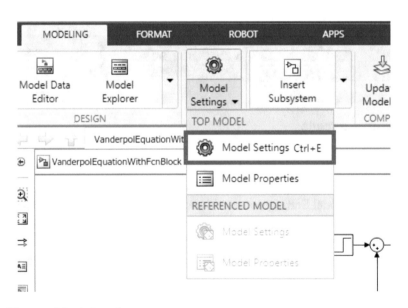

Figure 1.71: The model settings button.

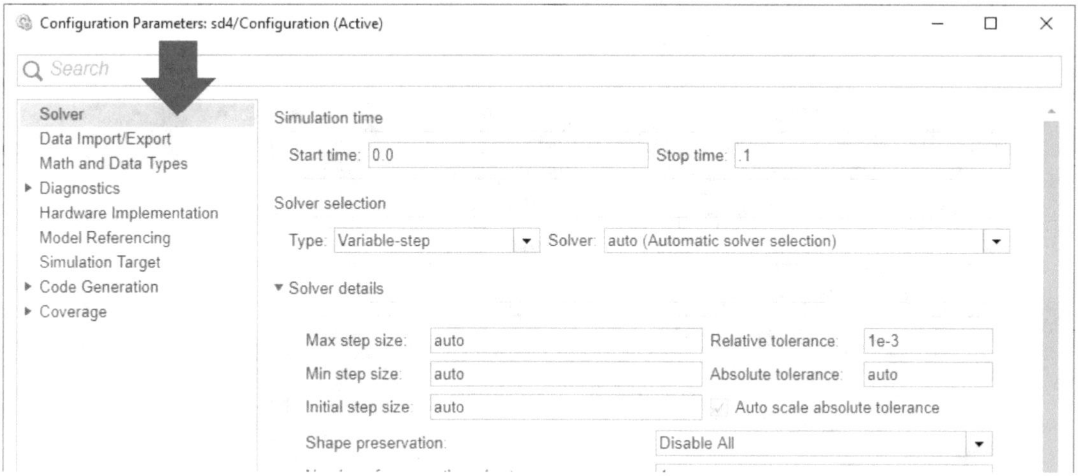

Figure 1.72: The solver section of configuration parameters window.

Figure 1.73: Decrease the number in the fixed-step size (fundamental sample time) box to get a smoother graph from Simulink.

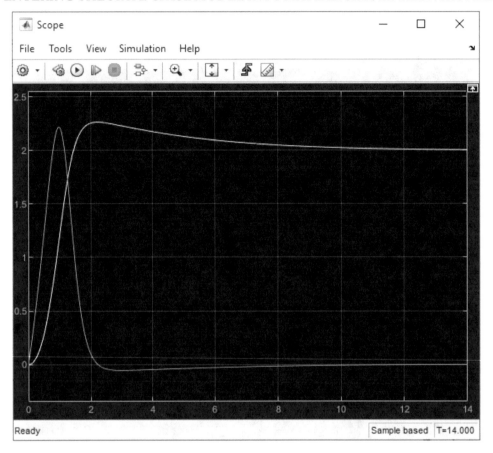

Figure 1.74: Smoothness of drawn curves are increased in comparison to Fig. 1.69.

As another example, we simulate the following equation:

$$
\begin{cases}
\dot{x} = v \\
\dot{\theta} = \omega \\
\dot{v} = \dfrac{F + \omega^2 \sin(\theta) - 9.8 \sin(\theta) \cos(\theta)}{2 - \cos(\theta)^2} \\
\dot{\omega} = \dfrac{-F \cos(\theta) - \omega^2 \sin(\theta) \cos(\theta) + 19.6 \sin(\theta)}{2 - \cos(\theta)^2}.
\end{cases}
\tag{1.23}
$$

This is the simplified dynamical equation of an inverted pendulum placed on a cart (Fig. 1.75 with $l = 1$ m, $m = M = 1$ kg). The cart has a small motor which produce the force F to move the cart and keep the pendulum in the upright position. The x is the position of the cart, θ is

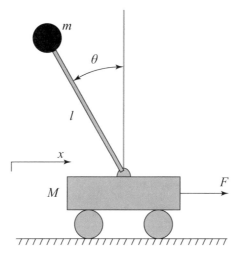

Figure 1.75: Inverted pendulum placed on a cart.

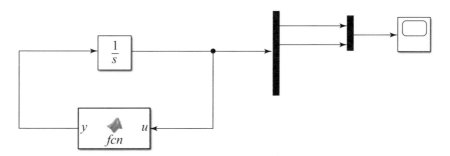

Figure 1.76: Simulation diagram for Equation (1.23).

the deflection of rod from the upright position, v is the speed of cart, and ω is the angular speed of the rod.

We want to simulate this nonlinear dynamical equation for $F = 0$ and the initial condition $(0,0.1,0,0)$, i.e., the pendulum is released from $\theta = 0$ without any initial speed.

The Simulink diagram is shown in Fig. 1.76. We are interested in the first two states (x and θ). The MATLAB function block and Integrator block settings are shown in Fig. 1.77 and 1.78, respectively.

After running the simulation, the result shown in Fig. 1.79 is obtained. This is expected, since if we release the inverted pendulum, the pendulum falls, i.e., the pendulum rotates toward right and the angle θ increases. However as the pendulum rotates, the cart move in the opposite direction, i.e., negative values of x.

```
 Editor -                                                              ⊙ ×
   MATLAB Function  ×  +
1      ⊟ function y = fcn(u)
2  –      y=[u(3);
3            u(4);
4            (u(4)^2*sin(u(2))-9.8*sin(u(2))*cos(u(2)))/(2-cos(u(2))^2);
5            (-u(4)^2*sin(u(2))*cos(u(2))+19.6*sin(u(2)))/(2-cos(u(2))^2)];
6
```

Figure 1.77: Inside of MATLAB function block.

Block Parameters: Integrator   ×

Integrator

Continuous-time integration of the input signal.

Parameters

External reset: none   ▼

Initial condition source: internal   ▼

Initial condition:

[0;0.1;0;0]   ⋮

☐ Limit output

☐ Wrap state

☐ Show saturation port

☐ Show state port

Absolute tolerance:

auto   ⋮

☐ Ignore limit and reset when linearizing

☑ Enable zero-crossing detection

State Name: (e.g., 'position')

[   ]

         OK   Cancel   Help   Apply

Figure 1.78: Settings of integrator block.

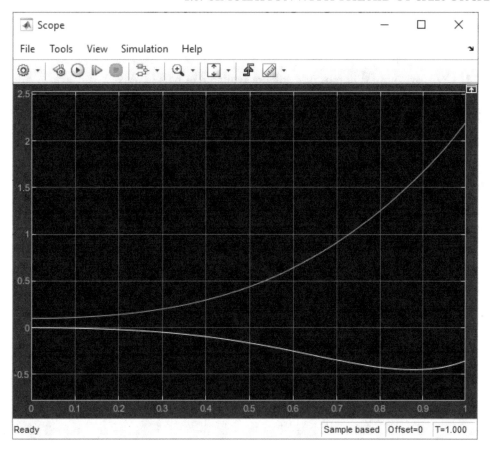

Figure 1.79: Result of simulation.

1.8   SIMULATION WITH THE AID OF ODE23 OR ODE45

In this section we simulate the Van der Pole equation (Equation (1.20) with $m = 1$) with the aid of ordinary differential solvers which is available in MATLAB. The Van der Pole Equation (1.20) is rewritten here for ease of reference:

$$\begin{cases} \dot{x}_1 = x_2 \\ \dot{x}_2 = m \left(1 - x_1^2 \right) x_2 - x_1. \end{cases} \tag{1.20}$$

MATLAB has many ode solvers: ode45, ode23, ode113, ode15s, ode23s, ode23t, ode23tb, and ode15i. Table 1.1 compares different solvers available in MATLAB.

The most commonly used solvers are ode23 and ode45. ode23 is a three-stage, third-order, Runge–Kutta method. ode45 is a six-stage, fifth-order, Runge–Kutta method. ode45 does more

Table 1.1: Comparison of different ode solvers available in MATLAB

Solver	Problem Type	Accuracy	When to use?
ode45	Nonstiff	Medium	Most of the time, ode45 should be the first solver you try.
ode23		Low	ode23 can be more efficient than ode45 at problems with crude tolerances, or in the presence of moderate stiffness.
ode113		Low to high	ode113 can be more efficient than ode45 at problems with stringent error tolerances, or when the ODE function is expensive to evaluate.
ode15s	Stiff	Low to medium	Try ode15s when ode45 fails or is inefficient and you suspect that the problem is stiff. Also use ode15s when solving differential algebraic equations (DAEs).
ode23s		Low	ode23s can be more efficient than ode15s at problems with crude error tolerances. It can solve some stiff problems for which ode15s is not effective. ode23s computes the Jacobian in each step, so it is beneficial to provide the Jacobian via odeset to maximize efficiency and accuracy. If there is a mass matrix, it must be constant.
ode23t		Low	Use ode23t if the problem is only moderately stiff and you need a solution without numerical damping. ode23t can solve differential algebraic equations (DAEs).
ode23tb		Low	Like ode23s, the ode23tb solver might be more efficient than ode15s at problems with crude error tolerances.
ode15i	Fully implicit	Low	Use ode15i for fully implicit problems $f(t,y,y') = 0$ and for differential algebraic equations (DAEs) of index 1.

Editor - D:\Books\ModernControl\m codes\vdpl.m

vdpl.m

```
1    function dydt = vdpl(t,x)
2
3    dydt = [x(2); (1-x(1)^2)*x(2)-x(1)];
```

Figure 1.80: Defining Equation (1.20).

Editor - D:\Books\ModernControl\m codes\vanderpoleExample.m

vdpl.m   vanderpoleExample.m

```
1    [t,x] = ode45(@vdpl,[0 14],[1; 2]);
2
3    figure(1)
4    plot(t,x(:,1),t,x(:,2))
5    title('Solution of van der Pol Equation (m = 1) with ODE45');
6    xlabel('Time t');
7    ylabel('x1 and x2');
8    legend('x_1','x_2');
9
10   figure(2)
11   plot(t,x(:,1))
12   xlabel('Time t(sec)');
13   ylabel('x1');
14
15   figure(3)
16   plot(x(:,1),x(:,2))
17   xlabel('x1');
18   ylabel('x2');
```

Figure 1.81: Solving the differential equations using the ode45 command.

work per step than ode23, but can take much larger steps. For differential equations with smooth solutions, ode45 is often more accurate than ode23.

We need to make a function for system dynamics before using the ode45 command. The function shown in Fig. 1.80 is the MATLAB translation of Van der Pole equation with $m = 1$. Use the code in Fig. 1.81 to solve the differential equation. This code solve the differential equation defined in the function for [0, 14] The initial condition of equation is (1,2).

After running the code, two new variables are added to the MATLAB Workspace. The variable t is the time points and the variable x is the states, i.e., x_1 and x_2 in Equation (1.20) (Fig. 1.82).

After running the code, the results shown in Figs. 1.83, 1.84, and 1.85 are obtained. This results are the same with the results produced by Simulink.

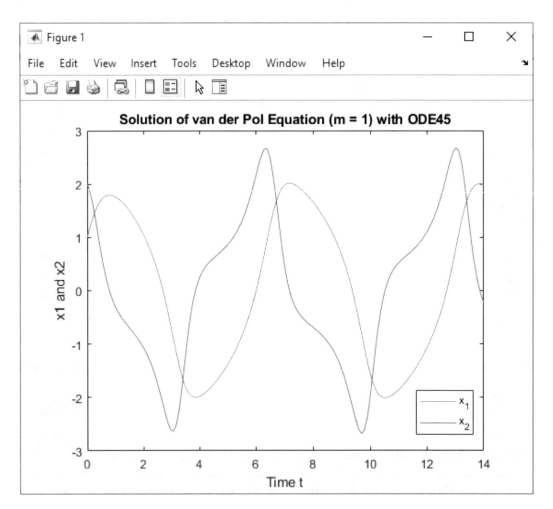

Figure 1.82: Addition of new variables to MATLAB workspace.

Figure 1.83: Result of simulation. This is the same curve with Fig. 1.60.

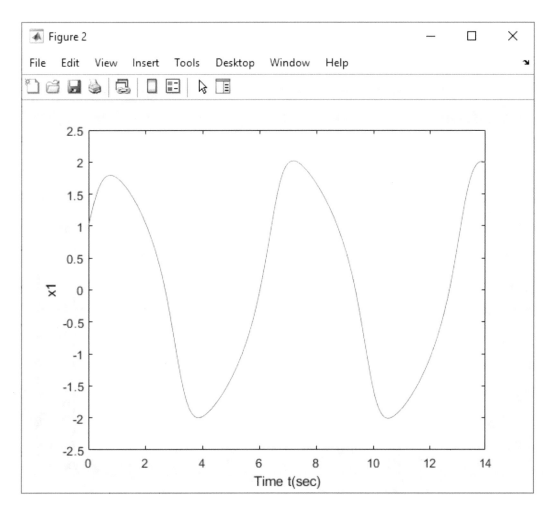

Figure 1.84: Output of system (variable x_1) vs. time. This is the same curve as in Fig. 1.64.

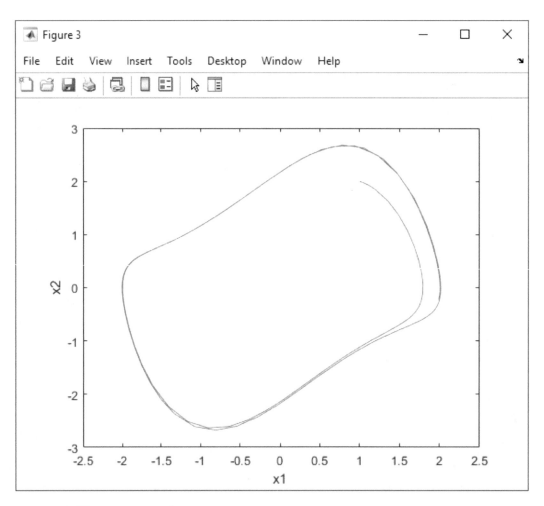

Figure 1.85: Phase portrait of system.

```
Editor - D:\Books\ModernControl\m codes\vanderpoleExample.m                    ⊙ ✕

   vdpl.m  ✕   vanderpoleExample.m  ✕   +

 1 -    [t,x] = ode45(@vdp1,[0 14],[1; 2]);
 2
 3 -    figure(1)
 4 -    plot(t,x(:,1),t,x(:,2))
 5 -    title('Solution of van der Pol Equation (m = 1) with ODE45');
 6 -    xlabel('Time t');
 7 -    ylabel('x1 and x2');
 8 -    legend('x_1','x_2');
 9
10 -    figure(2)
11 -    plot(t,x(:,1))
12 -    xlabel('Time t(sec)');
13 -    ylabel('x1');
14
15 -    figure(3)
16 -    plot(x(:,1),x(:,2))
17 -    xlabel('x1');
18 -    ylabel('x2');
19 -    axis([-10 10 -3 3])
```

Figure 1.86: Addition of axis command to code shown in Fig. 1.81.

Axis of the phase portrait in shown Fig. 1.62, are not in the same range as the one shown in Fig. 1.85. And because of this, with the naked eye, Fig. 1.85 may not look quite similar to the one shown in Fig. 1.62. Command axis must be used in order to obtain the phase portrait with the same axis range as the one shown in Fig. 1.62. Add line 19 to the code as shown in Fig. 1.86. The result of running the code shown in Fig. 1.86 is shown in Fig. 1.87. In this figure, range of axis is the same as the one shown in Fig. 1.62.

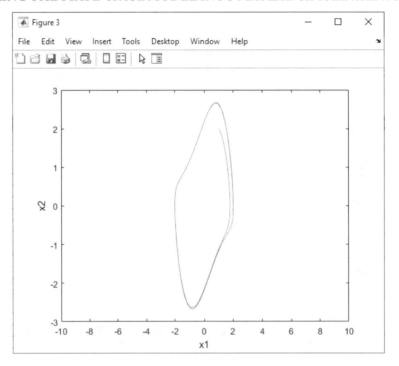

Figure 1.87: **Phase portrait of system.** This is the same curve as in Fig. 1.62.

1.9   FURTHER READING

[1] Jean Jacques E. Slotine and Weiping Li, *Applied Non-Linear Control*, Pearson, 1991. 31, 62

[2] Hassan K. Khalil, *Non-Linear Systems*, Pearson, 2001. 31, 62

[3] Katsuhiko Ogata, *State Space Analysis of Control System*, Prentice Hall, 1967. 63

[4] Bernard Friedland, *Control System Design: An Introduction to State-Space Methods*, Dover Publications Inc., 2005. 63

[5] Pierre R. Belanger, *Control Engineering: A Modern Approach*, Saunders College Publishing, 1995. 63

[6] Robert L. Williams II and Douglas A. Lawrence, *Linear State-Space Control Systems*, Wiley, 2007. 63

[1] and [2] are good references for nonlinear control. Definition of phase portrait and behavior around the equilibrium point could be found in the first chapter these references.

[3–6] are among the best references for linear state-space control systems and could be used to review the concepts when necessary.

CHAPTER 2

Linearization of Nonlinear Systems

2.1   INTRODUCTION

In this chapter we will study the problem of linearization of nonlinear state-space equations. Nearly all the systems in the real world are nonlinear. If you want to design a controller using linear techniques, then obtaining the linear model of system become important.

The Taylor series of a real function $f(x)$ that is infinitely differentiable at point a is the power series

$$f(a) + \frac{f'(a)}{1!}(x-a) + \frac{f''(a)}{2!}(x-a)^2 + \frac{f'''(a)}{3!}(x-a)^3$$
$$+ \ldots = \sum_{n=0}^{\infty} \frac{f^{(n)}(a)}{n!}(x-a)^n, \tag{2.1}$$

where $n!$ denotes the factorial of n. When $\Delta x \ll x$, we can say $f(x + \Delta x) \approx f(x) + f'(x) \times \Delta x$ in which prime shows the derivative with respect to variable x. The same rule is valid for multivariable functions as well. For instance, $f(x + \Delta x, y + \Delta y, z + \Delta z) \approx f(x, y, z) + \frac{\partial f}{\partial x}\Delta x + \frac{\partial f}{\partial y}\Delta y + \frac{\partial f}{\partial z}\Delta z$. Studying a numeric example is quite useful: assume $f(x, y, z) = x^4 y^3 + x^2 + \sqrt{y} + \frac{1}{z}$. Then $f(x + \Delta x, y + \Delta y, z + \Delta z) \approx x^4 y^3 + x^2 + \sqrt{y} + \frac{1}{z} + (4x^3 y^3 + 2x)\Delta x + (3x^4 y^2 + \frac{1}{2\sqrt{y}})\Delta y - \frac{1}{z^2}\Delta z$. Assume that we want to linearize this function around the point (1,2,3). Using the Taylor series, $f(1 + \Delta x, 2 + \Delta y, 3 + \Delta z) \approx 10.7475 + 34\Delta x + 12.354\Delta y - 0.111\Delta z$. So, the behavior of nonlinear function $f(x, y, z) = x^4 y^3 + x^2 + \sqrt{y} + \frac{1}{z}$ around the point (1,2,3) is estimated with the linear function $10.7475 + 34\Delta x + 12.354\Delta y - 0.111\Delta z$. For instance, $f(1.01, 1.94, 3.1) = 10.3352$. If we use the linear approximation, we obtain: $f(1.01, 1.94, 3.1) \approx 10.7475 + 34 \times 0.01 + 12.354 \times -0.06 - 0.111 \times 0.1 = 10.3352$ which is quite close to real value.

In engineering, some times notation $\delta x, \delta y, \delta z$, or $\tilde{x}, \tilde{y}$, and $\tilde{z}$ are used instead of $\Delta x, \Delta y$, and Δz. In this book, we use the tilde to show small signal perturbations around the linearization point. In this chapter, we linearize the systems using different methods. The first example is a solved using the pencil-and-paper analysis. This example shows the mathematical machinery of linearization. The second and third examples are solved using the MATLAB and Simulink, respectively. The results are the same despite of the method you use.

2.2   LINEARIZATION USING HAND CALCULATIONS

You can use the hand analysis for extraction of linear model of simple, low-order nonlinear systems. As our first example, we want to linearize the following nonlinear equation around the equilibrium point (40, 26.6667):

$$\begin{cases} \dot{x}_1 = -0.1x_1 - 0.01x_1^2 + 20 \\ \dot{x}_2 = 0.01x_1^2 - 0.6x_2. \end{cases} \tag{2.2}$$

It is quite easy to see that $x_1 = 40$ and $x_2 = 26.6667$ is the equilibrium of the system. Since $-0.1 \times 40 - 0.01 \times 40^2 + 20 = 0$ and $0.01 \times 40^2 - 0.6 \times 26.6667 = 0$.

You can calculate the equilibrium of given system with the aid of following code:

```
%Finds the equilibrium state of system
F = @(x) [-0.1*x(1)-0.01*x(1)^2+20;
          0.01*x(1)^2-0.6*x(2)];
x0=[10,20];

options =
optimoptions('fsolve','Display','iter');
[x,fval] = fsolve(F,x0,options)
```

After running the code, the result shown in Fig. 2.1 is obtained.

If you run the above code with x0=[-10,20], then you obtain the $(-50, 41.6667)$ which is another equilibrium point of the system (see Fig. 2.2). Here we linearize the system around the point (40, 26.6667). However, you can linearize the system around the $(-50, 41.6667)$ in the same way.

Value of x_1 around the 40, can be written as $x_1 = 40 + \tilde{x}_1 \cdot \tilde{x}_1$ shows the small signal perturbation around the operating point. In the same way, value of x_2 around the point 26.6667 can be written as $x_2 = 26.6667 + \tilde{x}_2$, in which $\tilde{x}_2$ shows the small signal perturbation of variable x_2.

We put this values in the original nonlinear equation and we use the first order Taylor series in order to linearize the nonlinear terms. Linear form of first state equation is:

$$\frac{d}{dt}(40 + \tilde{x}_1) = -0.1(40 + \tilde{x}_1) - 0.01(40^2 + 2 \times 40 \times \tilde{x}_1) + 20 \Rightarrow$$

$$\frac{d}{dt}(\tilde{x}_1) = -4 - 0.1\tilde{x}_1 - 16 - 0.8\tilde{x}_1 + 20 \Rightarrow \tag{2.3}$$

$$\frac{d}{dt}(\tilde{x}_1) = -0.9\tilde{x}_1.$$

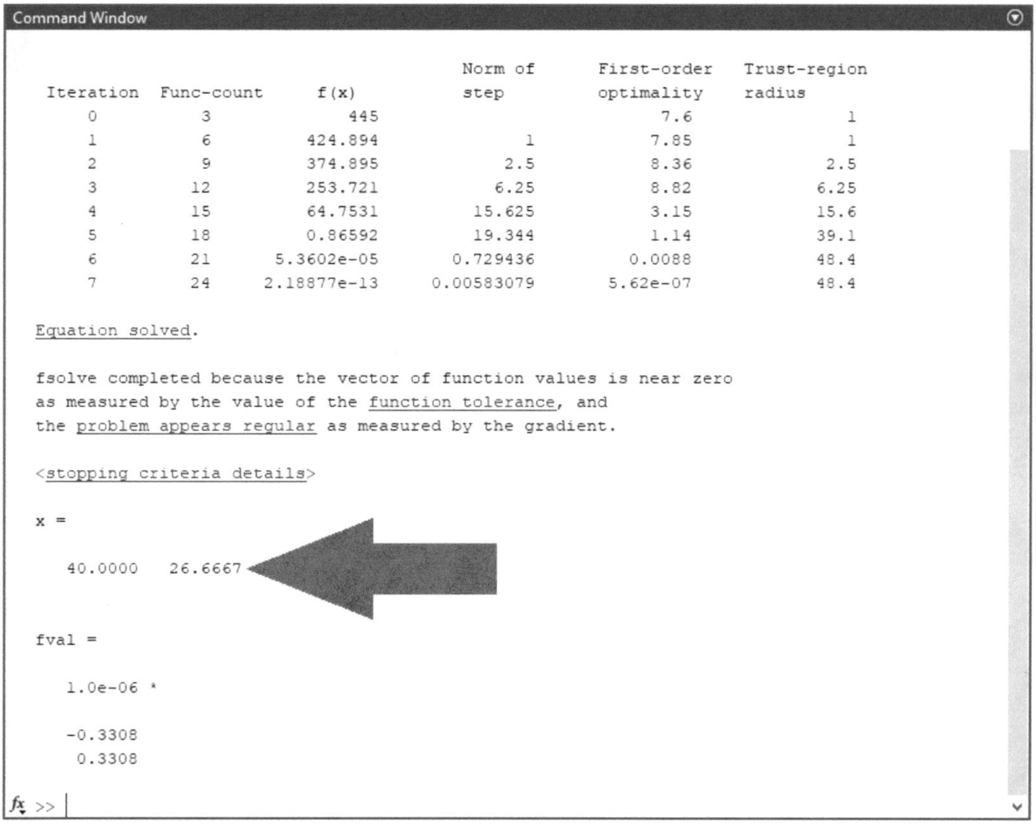

Figure 2.1: Calculating the equilibrium point of Equation (2.2) starting from point [10,20].

Note that we used the $(x + \Delta x)^2 \approx x^2 + 2 \times x \times \Delta x$ to linearize the nonlinear term in the first state equation. For instance $(10 + 0.2)^2 \approx 10^2 + 2 \times 10 \times 0.2 = 100 + 4 = 104$. This is a very good approximation since the correct value of $(10 + 0.2)^2 = 10.2^2 = 104.04$.

Linear form of second state equation is:

$$\dot{x}_2 = 0.01x_1^2 - 0.6x_2 \Rightarrow$$

$$\frac{d}{dt}(26.6667 + \tilde{x}_2) = 0.01(1600 + 2 \times 40 \times \tilde{x}_1) - 0.6(26.667 + \tilde{x}_2) \Rightarrow$$

$$\frac{d}{dt}(\tilde{x}_2) = 16 + 0.8\tilde{x}_1 - 16 - 0.6\tilde{x}_2 \Rightarrow \tag{2.4}$$

$$\frac{d}{dt}(\tilde{x}_2) = 0.8\tilde{x}_1 - 0.6\tilde{x}_2.$$

```
Command Window
                                     NOIM OI      IIIOU OIQCI     IIQOU IOYIOM
   Iteration   Func-count      f(x)        step       optimality    radius
       0           3            521                       6.6            1
       1           6          505.558          1          6.23           1
       2           9          468.607         2.5          5.3          2.5
       3          12          384.576        6.25         5.84         6.25
       4          15          227.169       15.625         3.5         15.6
       5          16          227.169       39.0625         3.5         39.1
       6          19          138.96        9.76563         3.8         9.77
       7          22          8.41201       24.4141         1.01        24.4
       8          25          0.0187087     7.63467        0.184         61
       9          28          2.64209e-08   0.108687      0.000218       61
      10          31          5.39778e-20   0.000129469   3.12e-10       61


Equation solved.

fsolve completed because the vector of function values is near zero
as measured by the value of the function tolerance, and
the problem appears regular as measured by the gradient.

<stopping criteria details>

x =

   -50.0000    41.6667


fval =

   1.0e-09 *

   -0.1645
    0.1641

fx >>
```

Figure 2.2: Calculating the equilibrium point of Equation (2.2) starting from point $[-10, 20]$.

So, the linearized system around the operating point (40, 26.6667) is:

$$
\begin{cases}
\dfrac{d}{dt}(\tilde{x}_1) = -0.9\tilde{x}_1 \\[2mm]
\dfrac{d}{dt}(\tilde{x}_2) = 0.8\tilde{x}_1 - 0.6\tilde{x}_2.
\end{cases}
\tag{2.5}
$$

Equation (2.5) can be written in matrix form as:

$$
\begin{bmatrix}
\dfrac{d}{dt}(\tilde{x}_1) \\[2mm]
\dfrac{d}{dt}(\tilde{x}_2)
\end{bmatrix}
=
\begin{bmatrix}
-0.9 & 0 \\
0.8 & -0.6
\end{bmatrix}
\begin{bmatrix}
\tilde{x}_1 \\
\tilde{x}_2
\end{bmatrix}.
\tag{2.6}
$$

We finished the hand analysis and its time to do solve the problem with MATLAB. We can linearize the given nonlinear system can be linearized with the aid of following code:

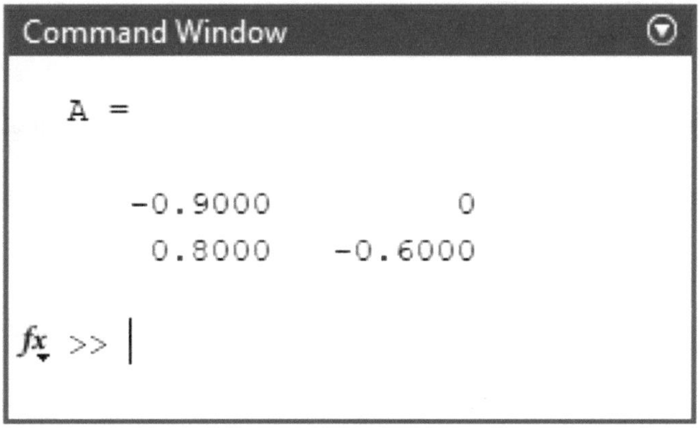

Figure 2.3: Value of matrix A.

```
%This code, linearize the system around the point (40,26.7)
syms x1 x2                %defining the symbolic variables
x0=[40 26.7];             %operating point

f1=-0.1*x1-0.01*x1^2+20;  %nonlinear state equation I
f2=0.01*x1^2-0.6*x2;      %nonlinear state equation II

% Linearization
a11=subs(diff(f1,x1),[x1 x2],x0);
a12=subs(diff(f1,x2),[x1 x2],x0);
a21=subs(diff(f2,x1),[x1 x2],x0);
a22=subs(diff(f2,x2),[x1 x2],x0);

A=eval([a11 a12;a21 a22]);
```

After running the code, the result shown in Fig. 2.3 is obtained. This is the same result that we obtained before in hand analysis.

2.3   LINEARIZATION OF INVERTED PENDULUM ON A CART USING MATLAB

As another example, we study the linearization problem of inverted pendulum on a cart. The dynamical equation of system in this example is:

$$\begin{cases} \dot{x} = v \\ \dot{\theta} = \omega \\ \dot{v} = \dfrac{F + \omega^2 \sin(\theta) - 9.8 \sin(\theta) \cos(\theta)}{2 - \cos(\theta)^2} \\ \dot{\omega} = \dfrac{-F \cos(\theta) - \omega^2 \sin(\theta) \cos(\theta) + 19.6 \sin(\theta)}{2 - \cos(\theta)^2}. \end{cases} \tag{2.7}$$

The following code linearize the system around the (0,0,0,0):

```
clc
clear all

%defining the symbolic variables (states)
%f1, f2, f3 and f4 are the nonlinear equations of system
%we want to linearize f1, f2, f3 and f4
         around the point (0,0,0,0)
syms x theta v w F
f1=v;
f2=w;
f3=(F+w^2*sin(theta)-9.8*sin(theta)*cos(theta))/(2-cos(theta)^2);
f4=(-F*cos(theta)-w^2*sin(theta)*cos(theta)
         +19.6*sin(theta))/(2-cos(theta)^2);

% op is the operating point of system.
         The system is linearized around this point.
op=[0,0,0,0];

% obtaining the matrix A
%the command diff is used to calculate the differentiation
         in the operating point.

%diff(f1,x) =df1/dx @ operating point op.
%diff(f1,theta) =df1/dtheta @ operating point op.
%diff(f1,v) =df1/dv @ operating point op.
```

```
%diff(f1,w) =df1/dw @ operating point op.
a11=subs(diff(f1,x),[x theta v w],op);
a12=subs(diff(f1,theta),[x theta v w],op);
a13=subs(diff(f1,v),[x theta v w],op);
a14=subs(diff(f1,w),[x theta v w],op);

%diff(f2,x) =df2/dx @ operating point op.
%diff(f2,theta) =df2/dtheta @ operating point op.
%diff(f2,v) =df2/dv @ operating point op.
%diff(f2,w) =df2/dw @ operating point op.
a21=subs(diff(f2,x),[x theta v w],op);
a22=subs(diff(f2,theta),[x theta v w],op);
a23=subs(diff(f2,v),[x theta v w],op);
a24=subs(diff(f2,w),[x theta v w],op);

%diff(f3,x) =df3/dx @ operating point op.
%diff(f3,theta) =df3/dtheta @ operating point op.
%diff(f3,v) =df3/dv @ operating point op.
%diff(f3,w) =df3/dw @ operating point op.
a31=subs(diff(f3,x),[x theta v w],op);
a32=subs(diff(f3,theta),[x theta v w],op);
a33=subs(diff(f3,v),[x theta v w],op);
a34=subs(diff(f3,w),[x theta v w],op);

%diff(41,x) =df4/dx @ operating point op.
%diff(f4,theta) =df4/dtheta @ operating point op.
%diff(f4,v) =df4/dv @ operating point op.
%diff(f4,w) =df4/dw @ operating point op.
a41=subs(diff(f4,x),[x theta v w],op);
a42=subs(diff(f4,theta),[x theta v w],op);
a43=subs(diff(f4,v),[x theta v w],op);
a44=subs(diff(f4,w),[x theta v w],op);

A=eval([a11 a12 a13 a14;a21 a22 a23 a24;a31 a32 a33 a34;
        a41 a42 a43 a44]);

%Obtaining the matrix B
%Here we differntiate with respect to F (external output).
b11=subs(diff(f1,F),[x theta v w],op);
```

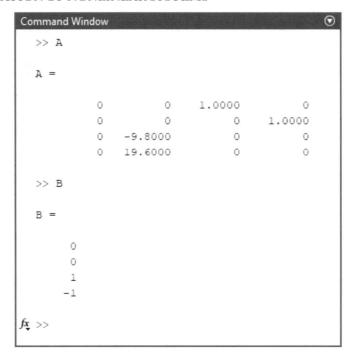

Figure 2.4: Value of matrices A and B.

```
b21=subs(diff(f2,F),[x theta v w],op);
b31=subs(diff(f3,F),[x theta v w],op);
b41=subs(diff(f4,F),[x theta v w],op);

B=eval([b11;b21;b31;b41]);
```

After running the code, the result shown in Fig. 2.4 is obtained. The linearized system is unstable since it has pole in Right Half Plane (RHP) (Fig. 2.5).

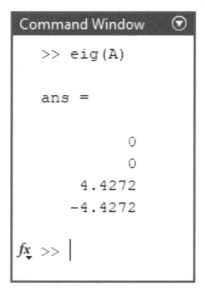

Figure 2.5: Eigen values of matrix A. Eigen values of matrix A are the system poles.

2.4 LINEARIZATION OF INVERTED PENDULUM ON A CART USING SIMULINK

In this section we linearize the inverted pendulum model with the aid of Simulink. The dynamical model inverted pendulum is rewritten below for easy reference:

$$
\begin{cases}
\dot{x} = v \\
\dot{\theta} = \omega \\
\dot{v} = \dfrac{F + \omega^2 \sin(\theta) - 9.8\sin(\theta)\cos(\theta)}{2 - \cos(\theta)^2} \\
\dot{\omega} = \dfrac{-F\cos(\theta) - \omega^2 \sin(\theta)\cos(\theta) + 19.6\sin(\theta)}{2 - \cos(\theta)^2}.
\end{cases}
\tag{2.7}
$$

The Simulink diagram of this equation is shown in Fig. 2.6. The input port is the input of the system, i.e., the force produced by the motor of cart. The input port can be found in the Commonly Used Blocks section of Simulink Library Browser (Fig. 2.7). The code inside the MATLAB function block is shown in Fig. 2.8. The settings of integrator block is shown in Fig. 2.9.

In the Simulink diagram, right click the line which connects the input port to the multiplexer. The color of line changes into blue (Fig. 2.10) and a menu appears (Fig. 2.11). Click the Linear Analysis Points.

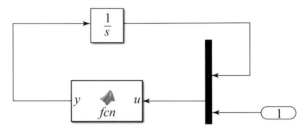

Figure 2.6: Simulink diagram of Equation (2.7).

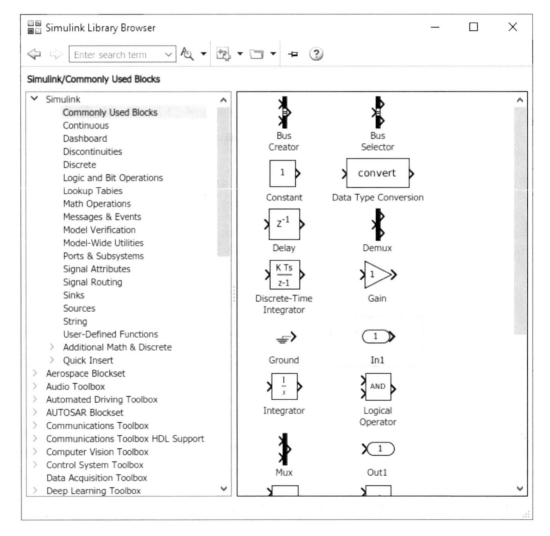

Figure 2.7: The input port block.

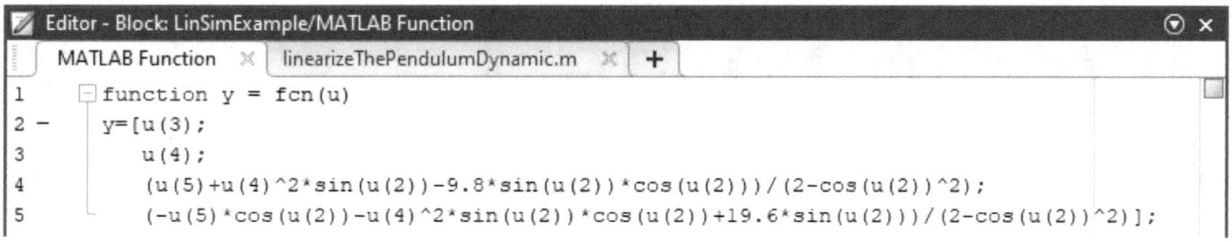

```
Editor - Block: LinSimExample/MATLAB Function                                                    ⊙ ✕
   MATLAB Function  ✕  linearizeThePendulumDynamic.m  ✕  +
1      ⊟function y = fcn(u)
2  −     y=[u(3);
3          u(4);
4          (u(5)+u(4)^2*sin(u(2))-9.8*sin(u(2))*cos(u(2)))/(2-cos(u(2))^2);
5          (-u(5)*cos(u(2))-u(4)^2*sin(u(2))*cos(u(2))+19.6*sin(u(2)))/(2-cos(u(2))^2)];
```

Figure 2.8: The code inside the MATLAB function block.

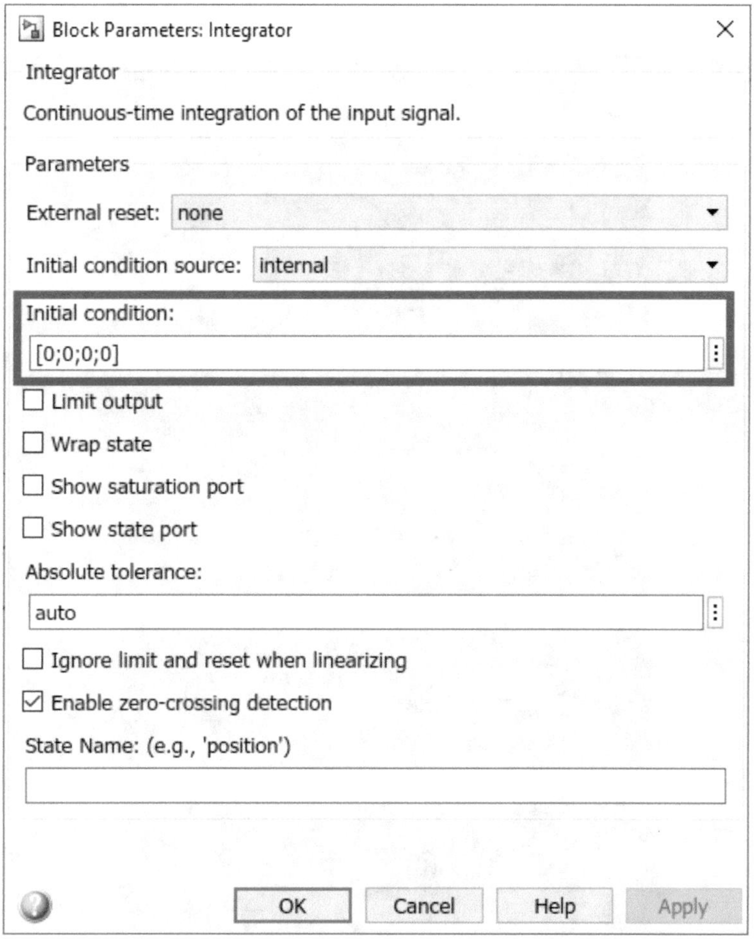

Figure 2.9: The integrator block settings.

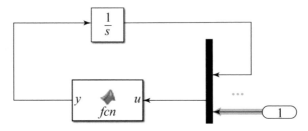

Figure 2.10: After right clicking the connection between input port and multiplexer, its color changes into blue.

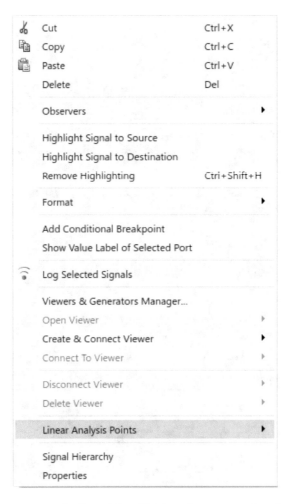

Figure 2.11: The menu appeared after right clicking the connection between input port and multiplexer.

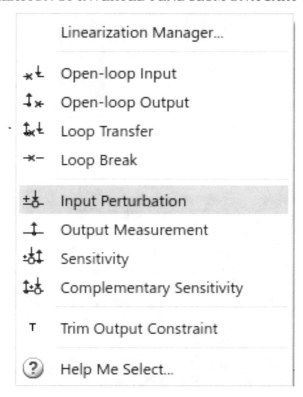

Figure 2.12: The menu appeared after clicking the linear analysis point. Select the input perturbation.

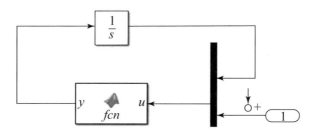

Figure 2.13: A small icon is added to the connection between input port and multiplexer.

After clicking the Linear Analysis Points, another menu will appear. Click the Input Perturbation (Fig. 2.12). After clicking the Input Perturbation, the Simulink model changes and a small icon will be added to the line which connects the input terminal to the multiplexer (Fig. 2.13).

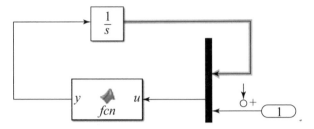

Figure 2.14: After right clicking the connection between integrator and multiplexer, its color changes into blue.

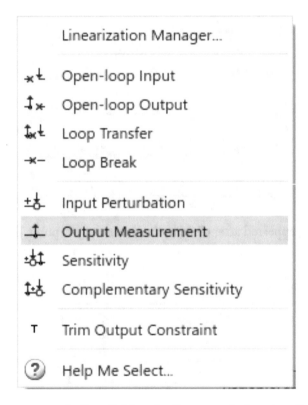

Figure 2.15: The menu appeared after clicking the linear analysis point. Select the output measurement.

We need to define the output of linearization as well. In order to do this, right click the line which connects the output of integrator to multiplexer (Fig. 2.14). Like the previous case, click the Linear Analysis Points from the appeared menu (Fig. 2.11). This time select the Output Measurement instead of Input Perturbation (Fig. 2.15).

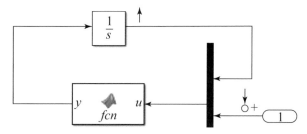

Figure 2.16: A small icon is added to the connection between integrator and multiplexer.

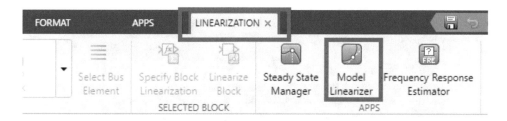

Figure 2.17: **LINEARIZATION** tab of Simulink.

After you clicked the Output Measurement, a small icon is added to the Simulink model (Fig. 2.16). Now, go to the **LINEARIZATION** tab. Click the Model Linearizer (Fig. 2.17).

Ensure that Operating Point: Model Initial Condition is selected (Fig. 2.18). We want the linearization to be done around (0,0,0,0) which is the initial condition defined in the integrator block. The (0,0,0,0) means that the pendulum stays in the upright position without any movement (with zero speed, pendulum angular speed = cart linear speed = 0).

In order to linearize the system, click the small triangle shown in Fig. 2.19. Click the Linearize button in the appeared menu (Fig. 2.20). After clicking the Linearize button, the Simulink linearize the model. Note that linear model appeared in the Linear Analysis Workspace section of Model Linearizer (Fig. 2.21).

You can linearize a nonlinear model and ask the Simulink to draw the step, Bode, impulse, Nyquist, or Nichols plot of linearized system for you after linearization. For instance, in order to see the step response of linear system, click on the Step button shown in Fig. 2.22. For Bode click the Bode button, etc.

The step response of linear system is shown in Fig. 2.23. We can deduce that system is unstable according to the drawn step response. Because the states increased without bound (the states reach to about 10^{24} in 14 s).

You can transfer the obtained linear model to MATLAB Workspace. To do this, simply drag and drop the obtained linear model from Linear Analysis Workspace to MATLAB Workspace (see Fig. 2.24).

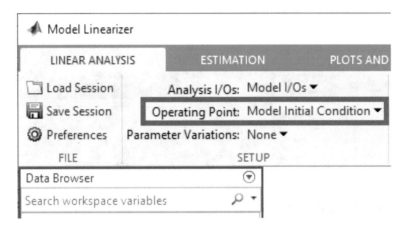

Figure 2.18: Ensure that model initial condition is selected.

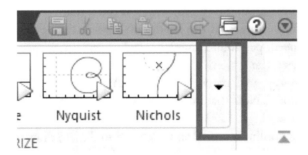

Figure 2.19: Click the small triangle in order to obtain access to linearize button.

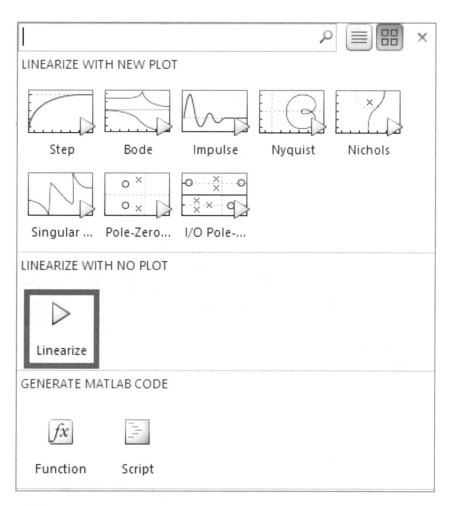

Figure 2.20: The linearize button.

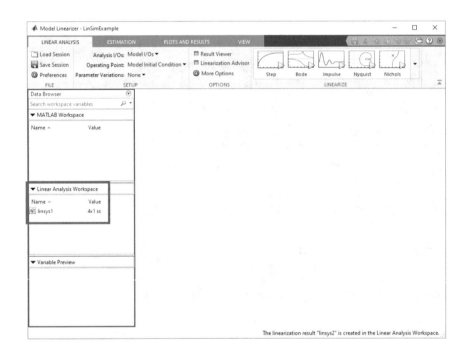

Figure 2.21: After linearization, a variable named linsys1 is added to linear analysis workspace. The linsys1 is the linearized model.

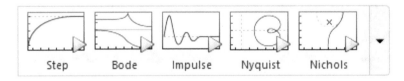

Figure 2.22: Step, bode, impulse, Nyquist, and Nichols buttons.

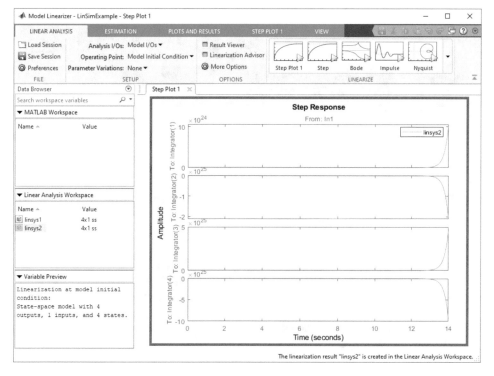

Figure 2.23: The step response of linearized system.

After drag and drop, a new variable will be added to MATLAB Workspace (Fig. 2.25). You can see the value of matrix A and B using the commands shown in Figs. 2.26 and 2.27, respectively. The matrices are the same with the one obtained before (see Fig. 2.4).

Let us take a look at system poles and see whether it is stable or not. This can be done with the aid of commands shown in Figs. 2.28 or 2.29. Since we have a pole in the RHP, the system is unstable. That is why the step response of the system increases with such a high speed (see Fig. 2.23).

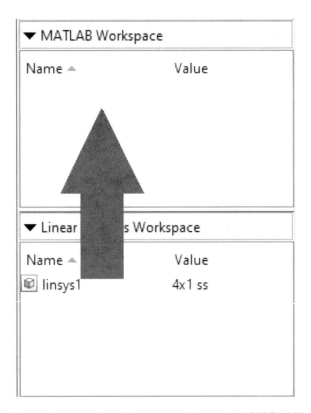

Figure 2.24: Drag and drop the calculated linear model to the MATLAB workspace.

Figure 2.25: The calculated linear model (linsys1) is transferred into the MATLAB workspace.

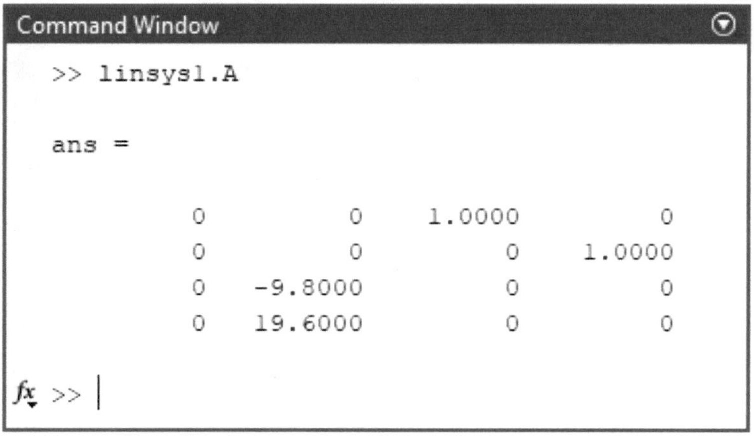

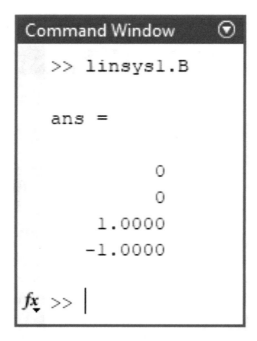

Figure 2.26: Matrix A of linsys1.

Figure 2.27: Matrix B of linsys1.

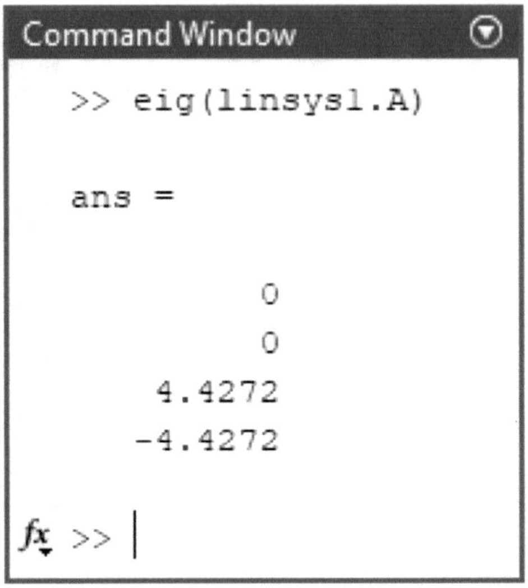

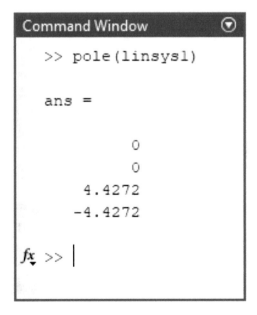

Figure 2.28: Eigen values of `linsys1`.

Figure 2.29: Poles of `linsys1`.

2.5   FURTHER READING

[1] C. Henry Edward and David E. Penney, *Calculus*, Pearson, 2002. 87

[2] Farzin Asadi and Kei Eguchi, *Dynamics and Control of DC-DC Converters*, Morgan & Claypool publishers, 2019. DOI: 10.2200/s00828ed1v01y201802pel010. 87

[1] is a very good calculus book. It could be used to review the Taylor series if you need.

[2] has many examples of linearizing the nonlinear dynamic equation of DC-DC converters. Refer to it for further examples of linearization.

CHAPTER 3

Designing State Feedback Controller

3.1 INTRODUCTION

The first chapter of this book studied the tools for modeling and analysis of systems. The second chapter studied the problem of linearization of nonlinear systems, which is quite important since nearly all the systems in the real world are nonlinear. Without linearization and a linear model, we cannot apply the linear techniques, which are in general simple and cheaper to implement in comparison to nonlinear techniques.

In this chapter, we will study the problem of designing controllers using the state-space techniques. The systems studied in this chapter are taken from [1]. It is a good idea to take a look to your textbook in order to remember the concepts of controllability and observability of linear system. In the first two examples, full state feedback is used. Full-state feedback needs as many sensors as the system states, which is not practical in many real-world problems.

If full-state feedback is not possible, we could use the observer to *estimate* the states. Good estimations, i.e., the one which converges to the real values of states in a reasonably fast time, could be used instead of real values to produce the control signal which applies to the plant. The third example uses the observer-based controller.

3.2 EXAMPLE 1: DC MOTOR SPEED CONTROL

DC motors are one of the most used actuators used in control systems. In this example, we study the problem of controlling the speed of a DC motor.

3.2.1 OBTAINING THE STATE-SPACE MODEL OF SYSTEM

The electric equivalent circuit of the armature and the free-body diagram of the rotor are shown in Fig. 3.1. In Fig. 3.1, J is moment of inertia of the rotor, b is motor viscous friction constant, R is the armature resistance, and L is the armature inductance.

The torque generated by a DC motor depends on two factors: the armature current and strength of magnetic field around armature. We assume that the strength of magnetic field around the armature is constant, i.e., the field is provided by a piece of permanent magnets. In this case the torque is proportional to only armature current. Such a motor is called armature-

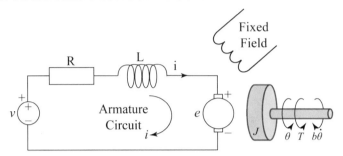

Figure 3.1: Model of DC motor.

controlled motor:

$$T = K_t i. \tag{3.1}$$

The motor back emf (electro motive force) is proportional to the angular velocity of shaft

$$e = K_e \dot{\theta} = K_e \omega. \tag{3.2}$$

The motor torque and back emf constants are equal, that is,

$$K_t = K_e = K. \tag{3.3}$$

From Fig. 3.1, we can derive the following governing equations based on Newton's 2nd law and Kirchhoff's voltage law:

$$\begin{aligned} J\ddot{\theta} + b\dot{\theta} &= Ki \\ L\frac{di}{dt} + Ri &= V - K\dot{\theta}. \end{aligned} \tag{3.4}$$

We can write the following state-space model for these equations:

$$\frac{d}{dt}\begin{bmatrix} \dot{\theta} \\ i \end{bmatrix} = \begin{bmatrix} -\dfrac{b}{J} & \dfrac{K}{J} \\ -\dfrac{K}{L} & -\dfrac{R}{L} \end{bmatrix}\begin{bmatrix} \dot{\theta} \\ i \end{bmatrix} + \begin{bmatrix} 0 \\ \dfrac{1}{L} \end{bmatrix}V$$

$$y = [1\ 0]\begin{bmatrix} \dot{\theta} \\ i \end{bmatrix}. \tag{3.5}$$

In this model, the armature voltage is treated as the input and the rotational speed is chosen as the output.

Table 3.1: DC motor parameters

Parameter	Value
J	0.01 kg.m^2
B	0.1 N.m.s
Ke	0.01 V.s/rad
Kt	0.01 N.m/A
R	1 Ω
L	0.5 H

You can use the ω instead of $\dot{\theta}$ if you prefer, i.e.,

$$\frac{d}{dt}\begin{bmatrix} \omega \\ i \end{bmatrix} = \begin{bmatrix} -\dfrac{b}{J} & \dfrac{K}{J} \\ -\dfrac{K}{L} & -\dfrac{R}{L} \end{bmatrix}\begin{bmatrix} \omega \\ i \end{bmatrix} + \begin{bmatrix} 0 \\ \dfrac{1}{L} \end{bmatrix} V$$

$$y = [1\ 0]\begin{bmatrix} \omega \\ i \end{bmatrix}.$$

(3.6)

The state-space model of motor is of the form:

$$\begin{cases} \dot{x} = Ax + Bu \\ y = Cx. \end{cases}$$

(3.7)

Assume a motor with the parameters as given by Table 3.1.
You can make the state-space model of this system with the aid of following codes:

```
J = 0.01;
b = 0.1;
K = 0.01;
R = 1;
L = 0.5;
A = [-b/J    K/J
     -K/L    -R/L];
B = [0
     1/L];
C = [1    0];
D = 0;
motor_ss = ss(A,B,C,D)
```

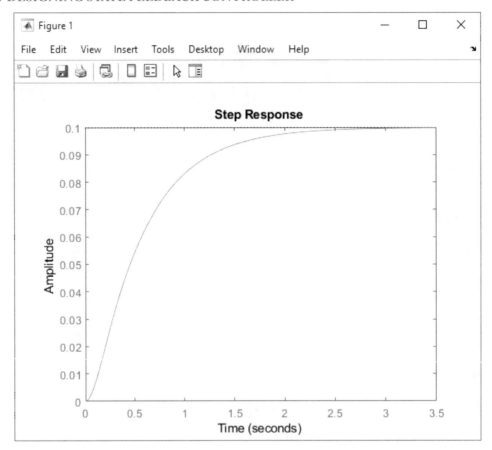

Figure 3.2: Step response of DC motor with parameter values given as given in Table 3.1.

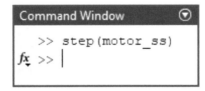

Figure 3.3: `step` command is used for drawing the step response of DC motor.

3.2.2   DESIGNING THE FULL-STATE FEEDBACK CONTROLLER

The step response of system is shown in Fig. 3.2. The step response is obtained with the aid command shown in Fig. 3.3.

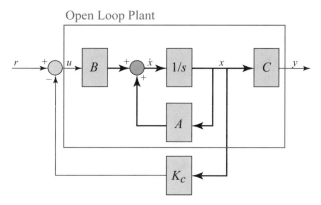

Figure 3.4: System controlled with the full-state feedback.

According to Fig. 3.2, the settling time is about 3 s and the system cannot follow the step command with zero steady-state error. So, there is a need to design a controller.

Assume that for a 1 rad/s step reference, the design criteria are the following.

1. Settling time less than 2 s.

2. Overshoot less than 5%.

3. Steady-stage error less than 1%.

Assume we can measure both of the states, i.e., we have a tachometer to measure speed of armature and a current sensor to measure the armature current. In this case we can design a full-state feedback controller and there is no need for observer.

The control law for a full-state feedback system has the form $u = r - K_c x$ and the associated schematic is shown in Fig. 3.4. By full-state, we mean that all state variables are known to the controller at all times.

If the given system is controllable, then by designing a full-state feedback controller we can move these two poles anywhere we'd like. Whether the given system is controllable or not can be determined by checking the rank of the controllability matrix $[B\ AB\ A^2B\ \ldots]$.

The MATLAB command ctrb constructs the controllability matrix given matrices A and B. Additionally, the command rank determines the rank of a given matrix. The commands shown in Fig. 3.5 verify the system's order and whether or not the system is controllable.

According to results shown in Fig. 3.5, we know that our system is controllable since the controllability matrix is full rank.

Let us put the closed-loop poles at -5+i and -5-i (note that this gives close to 0% overshoot and 0.8 s settling time). Once we have determined the pole locations we desire, we can use the MATLAB commands place or acker to determine the controller gain matrix, K_c, to achieve these poles.

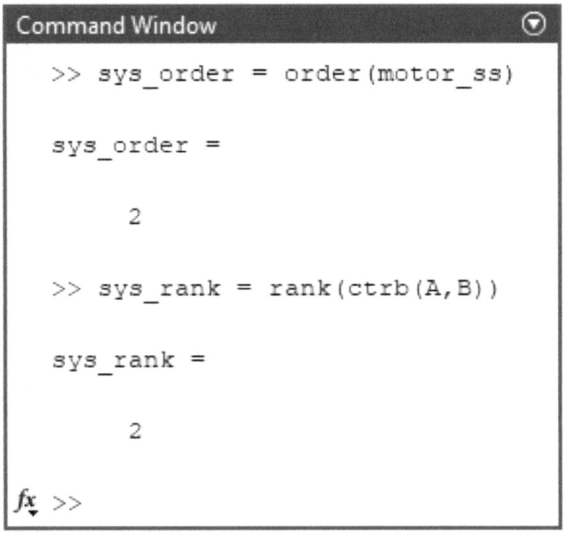

```
Command Window                              ⊙
    >> sys_order = order(motor_ss)

    sys_order =

         2

    >> sys_rank = rank(ctrb(A,B))

    sys_rank =

         2

fx >>
```

Figure 3.5: Calculation of system order and rank of controllability matrix.

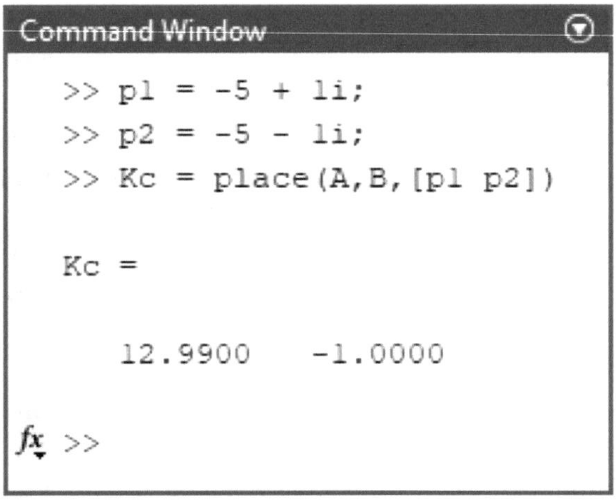

```
Command Window                              ⊙
    >> p1 = -5 + 1i;
    >> p2 = -5 - 1i;
    >> Kc = place(A,B,[p1 p2])

    Kc =

         12.9900    -1.0000

fx >>
```

Figure 3.6: Matrix of required gain.

Use of the command `place` is suggested since it is numerically better conditioned than acker. However, if we wished to place a pole with multiplicity greater than the rank of the matrix B, then we would have to use the command `acker`. For the code shown in Fig. 3.6, calculate the required gains.

```
Command Window                                              ⊙
    >> t = 0:0.01:3;
    >> sys_cl = ss(A-B*Kc,B,C,D);
    >> step(sys_cl,t)
    >> title('Step Response with State-Feedback Controller')
fx >> |
```

Figure 3.7: Commands required for drawing the step response of closed-loop system.

The state-space model of motor is of the form:

$$\begin{cases} \dot{x} = Ax + Bu \\ y = Cx. \end{cases} \tag{3.8}$$

If we substitute the state-feedback law $u = r - K_c x$ for u, we obtain the following expression:

$$\begin{cases} \dot{x} = (A - BK_c) x + Br \\ y = Cx. \end{cases} \tag{3.9}$$

We can then see the closed-loop response with the aid of commands shown in Fig. 3.7. The result of these commands are shown in Fig. 3.8.

In order to know the settling time of closed-loop system, right click on the plot. From the appeared menu, select the Characteristics. Select the Settling Time from the appeared menu (Fig. 3.9). The closed-loop system has settling time of 1.1 s (Fig. 3.10).

Although the settling time of a closed-loop system is in the acceptable range, there is a steady-state error which is not acceptable according to the given specification. Let us measure the steady-state value of drawn plot. In order to do this, right click on the plot, select Characteristics and then steady-state (Fig. 3.11). After clicking the steady-state, MATLAB finds the steady-state value of drawn plot for you. According to Fig. 3.12, the steady-state value of step response is 0.0769.

3.2.3   ADDITION OF A GAIN BLOCK FOR SOLVING THE STEADY-STATE ERROR PROBLEM

In Fig. 3.13, we added a gain block $\overline{N}$ with the value of $\frac{1}{0.0769} = 13.0039$ to the system. Step response of this system is shown in Fig. 3.14. This response satisfies the given specification.

You can solve the problem of steady-state errors, with the aid of integral control, as well. We will study this method in the next example of this chapter.

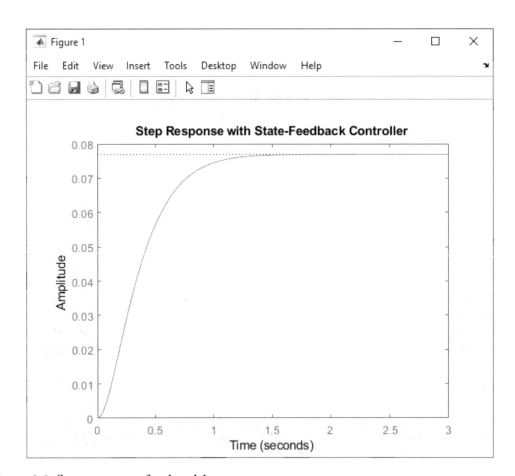

Figure 3.8: Step response of a closed-loop system.

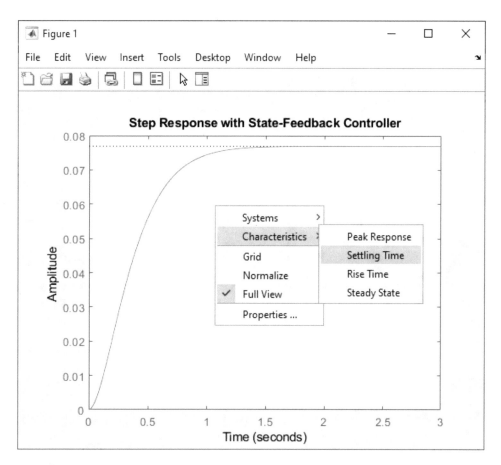

Figure 3.9: Select the settling time in order to see the settling time of the system.

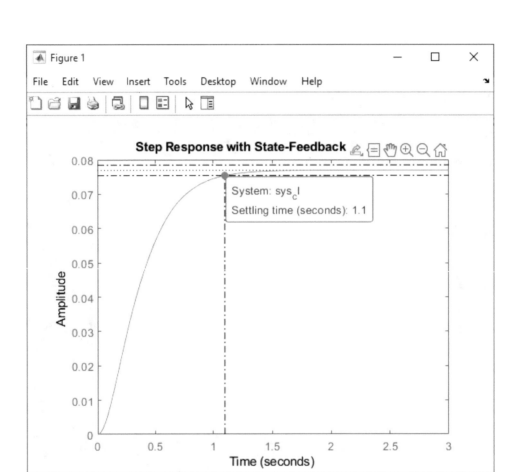

Figure 3.10: MATLAB calculates the settling time. It is 1.1 s.

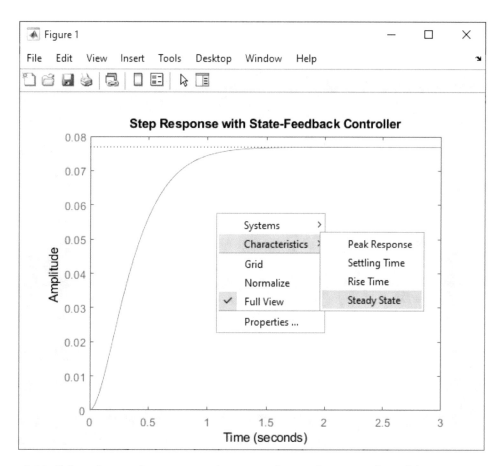

Figure 3.11: Select the steady-state in order to see the steady-state value of the system.

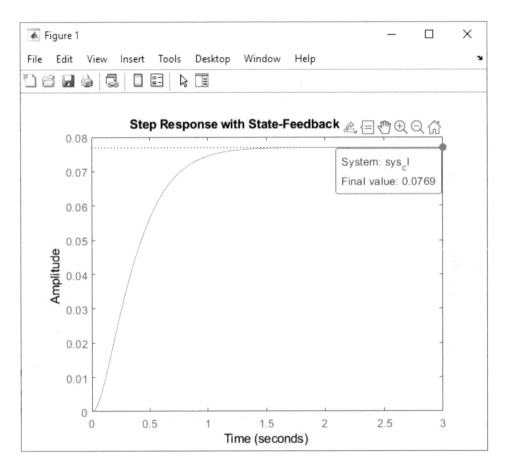

Figure 3.12: MATLAB calculates the steady-state value. It is 76.9 ms.

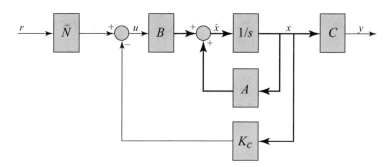

Figure 3.13: Addition of block $\overline{N}$ in order to solve the problem of steady-state error.

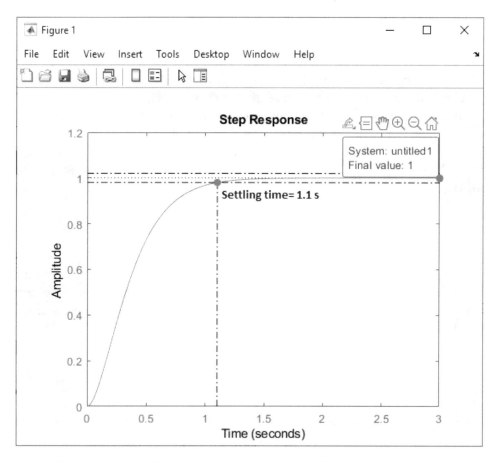

Figure 3.14: Step response of system shown in Fig. 3.13. The settling time is 1.1 s and the steady-state error problem is solved.

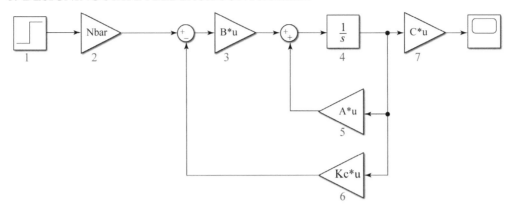

Figure 3.15: The Simulink diagram of a closed-loop system with state feedback.

Let us simulate the designed control system in Simulink. We use the same blocks as shown in Fig. 3.15. The settings of used blocks are shown in Figs. 3.16–3.22. The simulation result is shown in Fig. 3.23. This is the same result as the one shown in Fig. 3.14.

You can use the state-space block in order to simulate the closed-loop system as well. This method is shown in Fig. 3.24. The setting of Step block and Nbar gain block are the same as the one shown in Figs. 3.16 and 3.17. The setting of state-space block and Gain blocks of Fig. 3.24 are shown in Figs. 3.25–3.27. The simulation result is shown in Fig. 3.28. As seen, the result is the same with Figs. 3.14 and 3.23.

3.3   EXAMPLE 2: DC MOTOR POSITION CONTROL PROBLEM

In this example, we design a full-state feedback controller for a rotational position control system. The actuator is a DC motor. The state variables are armature rotational position, armature speed, and armature current. Again, the armature voltage is treated as the input and the rotational position is chosen as the output.

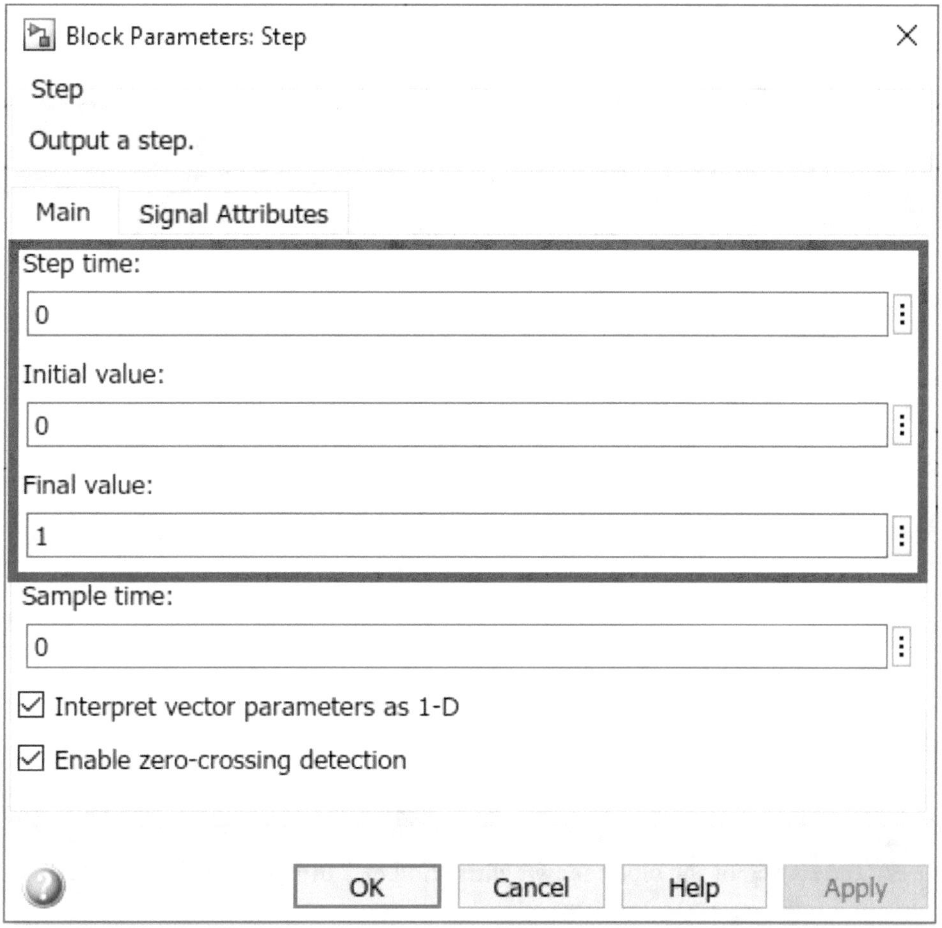

Figure 3.16: The setting of the block shown with number 1 in Fig. 3.15.

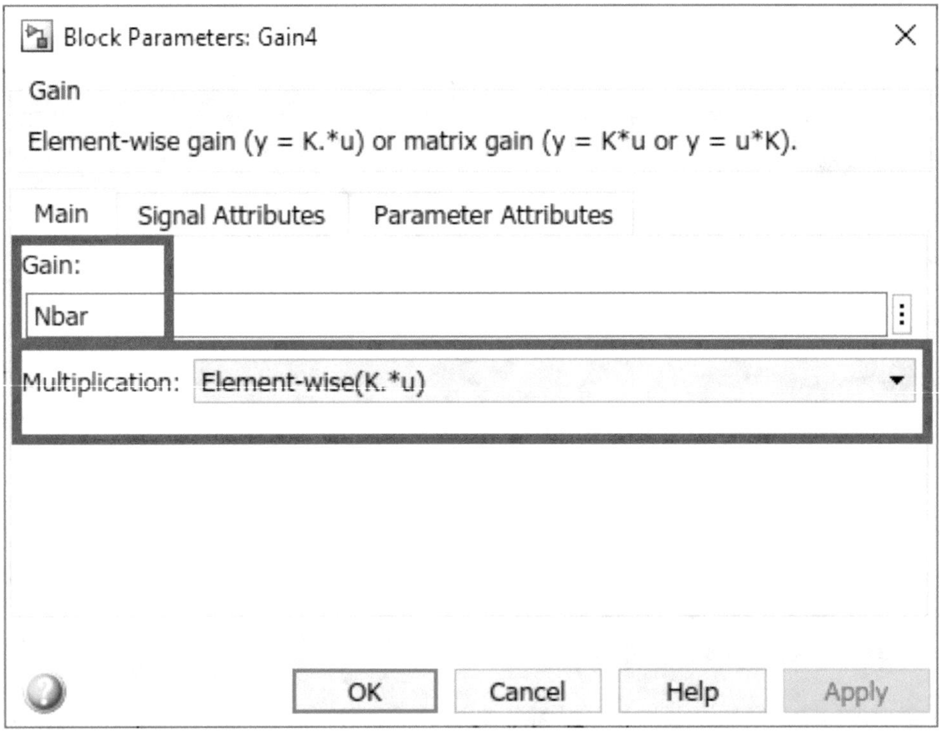

Figure 3.17: The setting of the block shown with number 2 in Fig. 3.15 ($\overline{N} = 13.0039$).

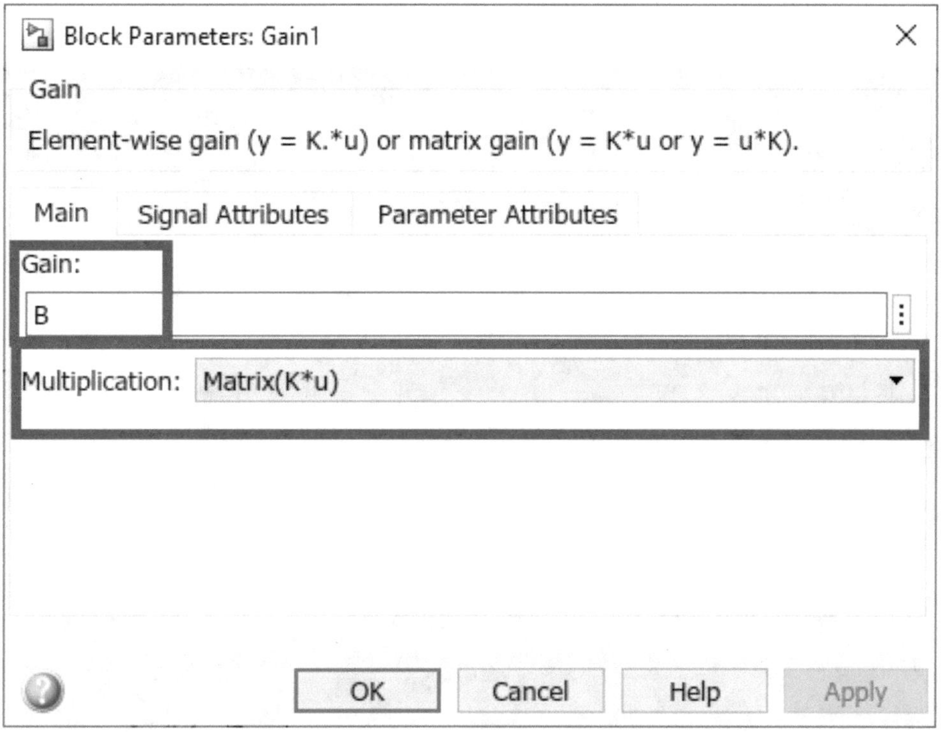

Figure 3.18: The setting of the block shown with number 3 in Fig. 3.15.

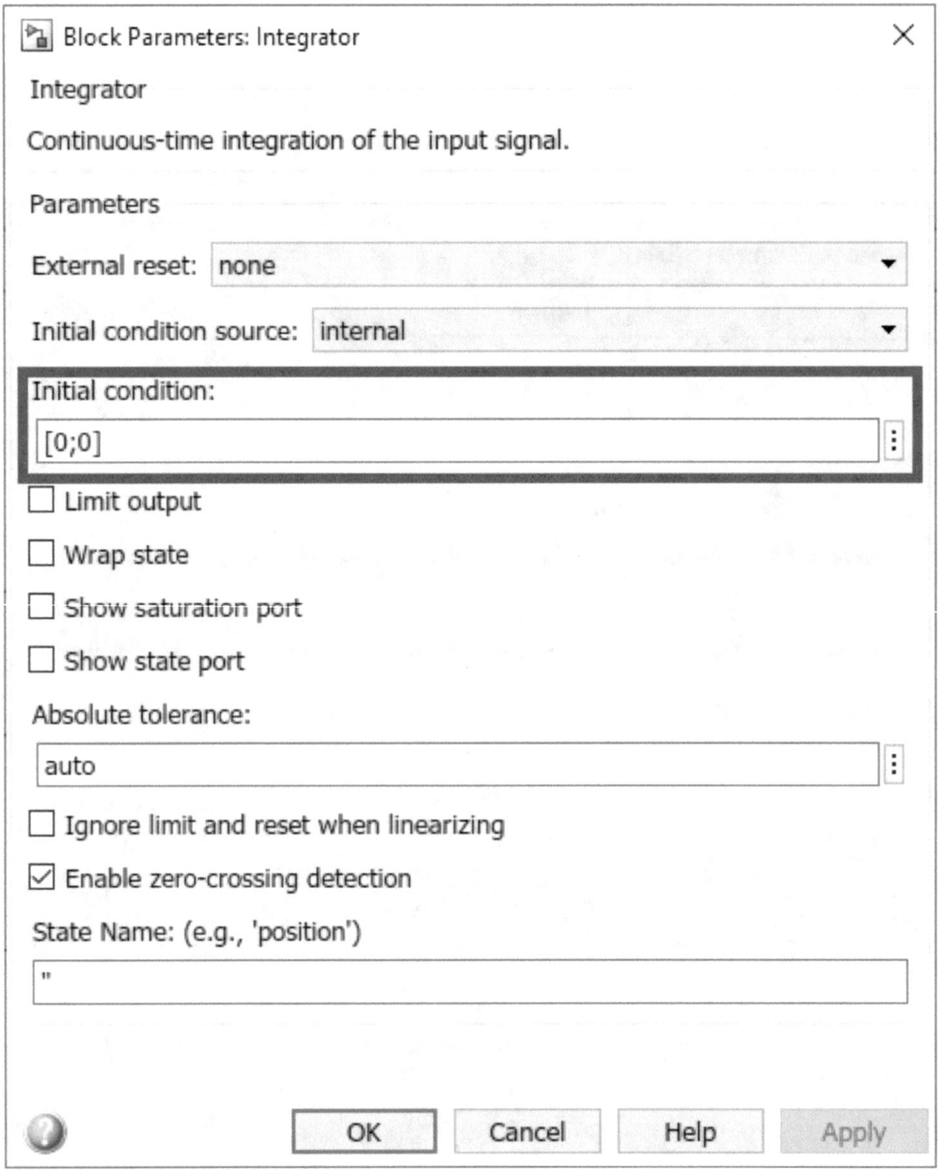

Figure 3.19: The setting of the block shown with number 4 in Fig. 3.15.

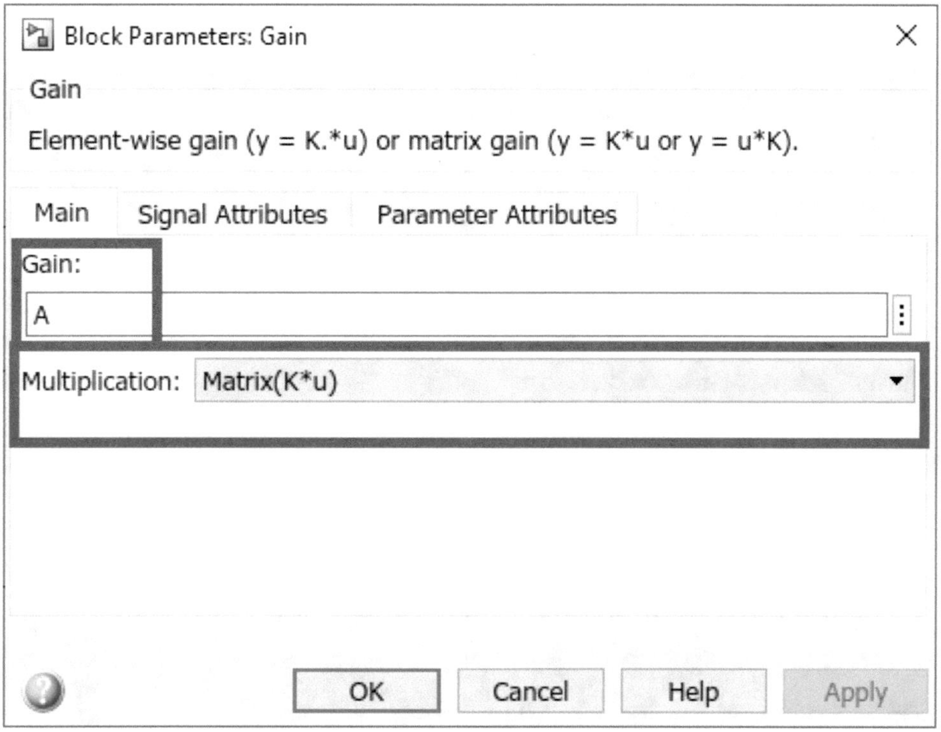

Figure 3.20: The setting of the block shown with number 5 in Fig. 3.15.

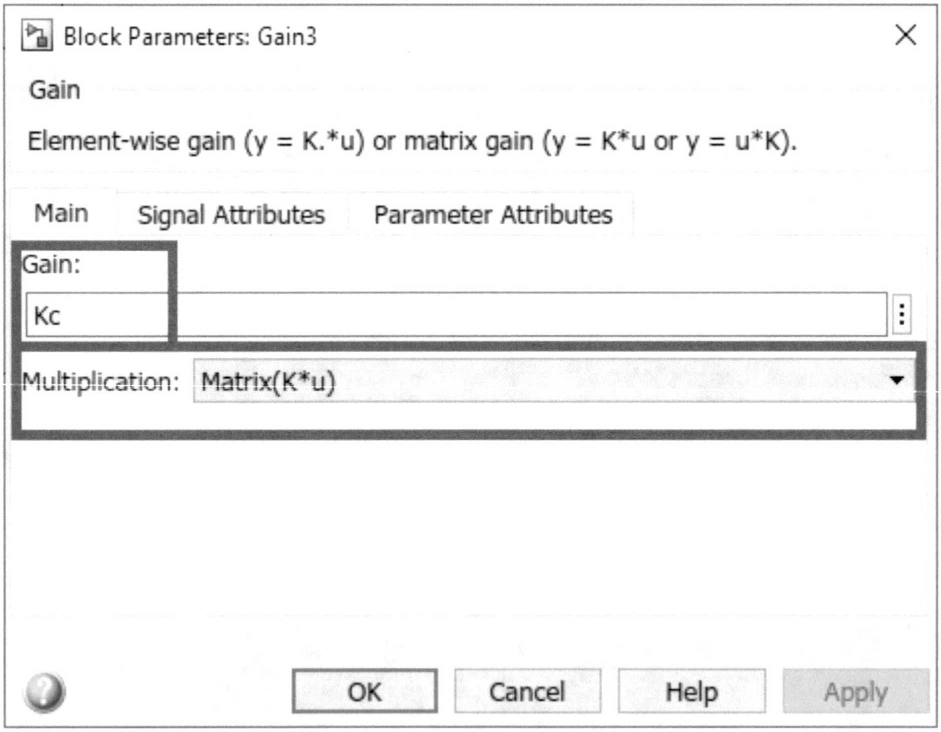

Figure 3.21: The setting of the block shown with number 6 in Fig. 3.15.

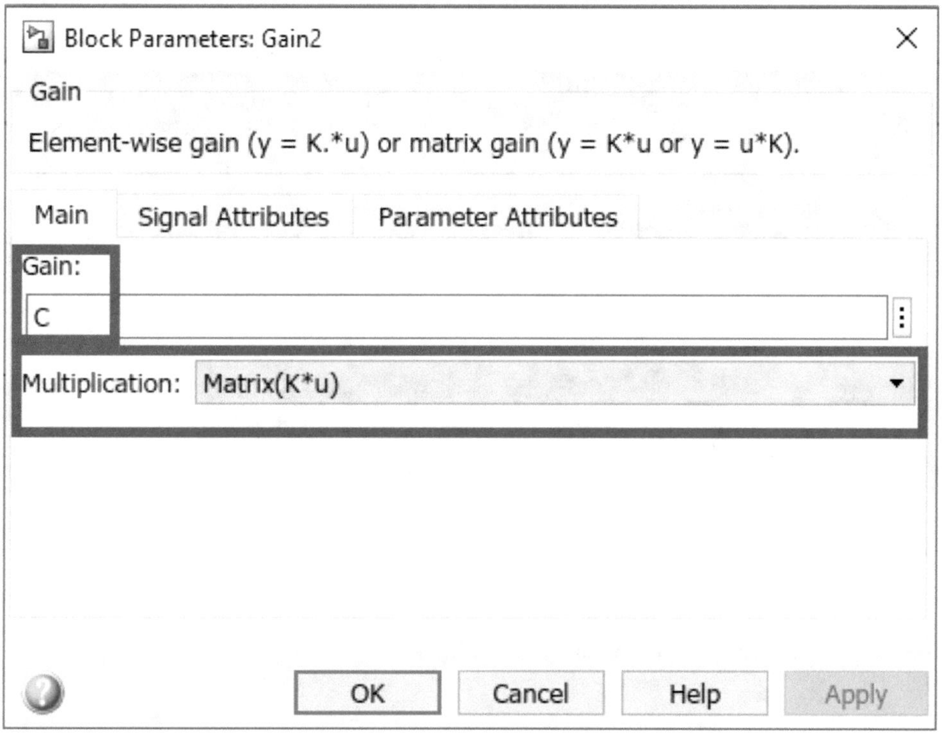

Figure 3.22: The setting of the block shown with number 7 in Fig. 3.15.

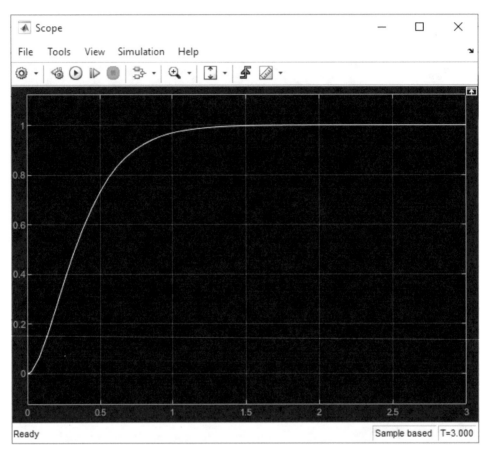

Figure 3.23: Step response of system shown in Fig. 3.15.

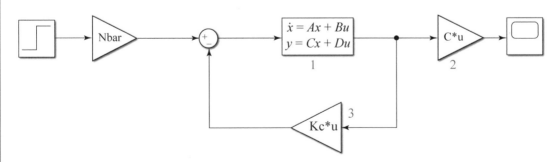

Figure 3.24: Simulation of the system with the aid of state-space block.

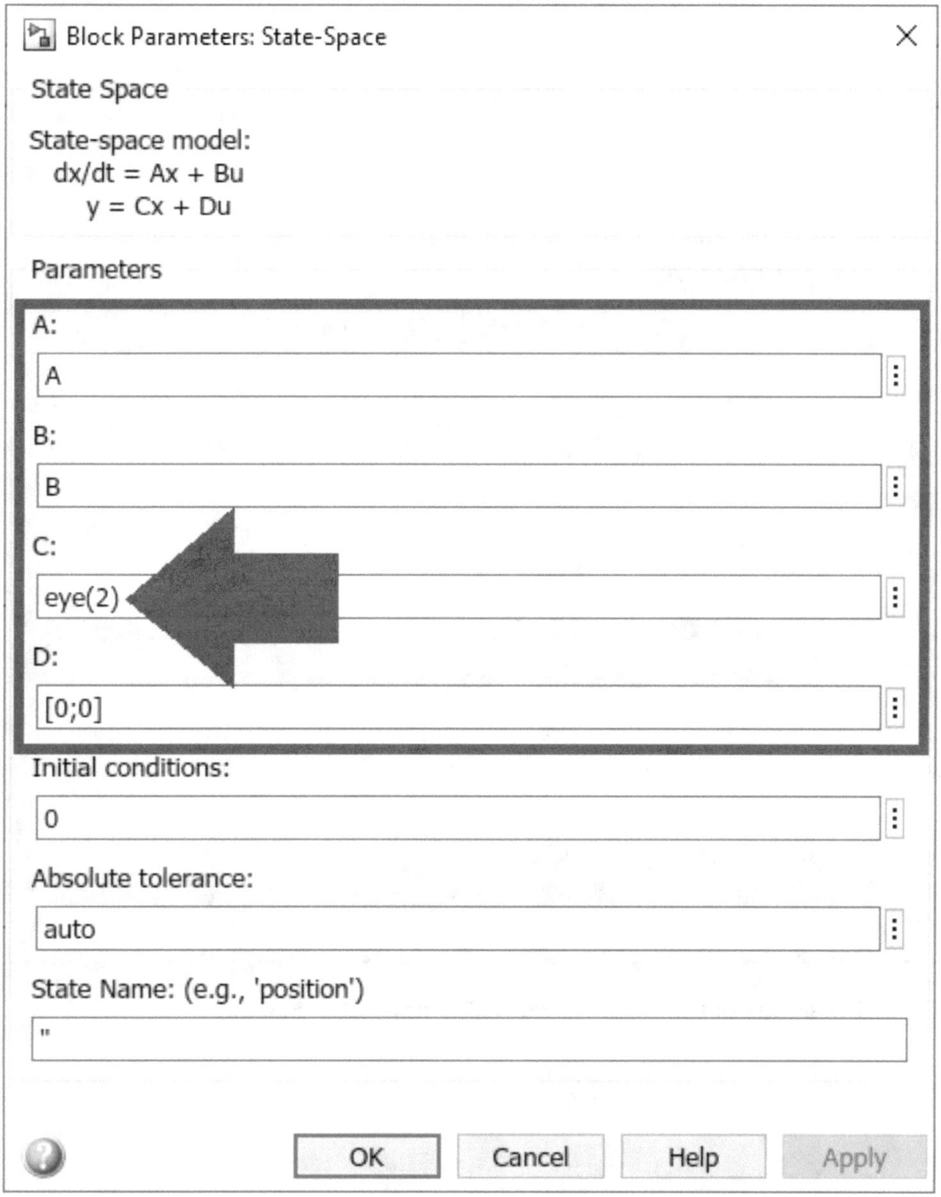

Block Parameters: State-Space   ✕

State Space

State-space model:
   dx/dt = Ax + Bu
       y = Cx + Du

Parameters

A:

A

B:

B

C:

eye(2)

D:

[0;0]

Initial conditions:

0

Absolute tolerance:

auto

State Name: (e.g., 'position')

"

OK   Cancel   Help   Apply

Figure 3.25: The setting of the block shown with number 1 in Fig. 3.24.

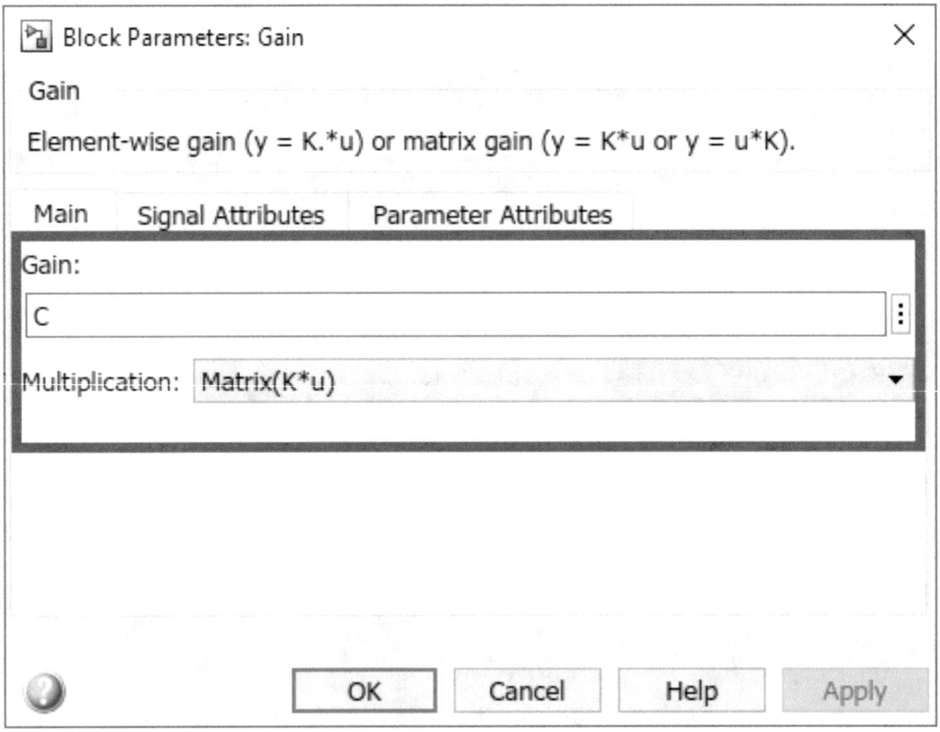

Figure 3.26: The setting of the block shown with number 2 in Fig. 3.24.

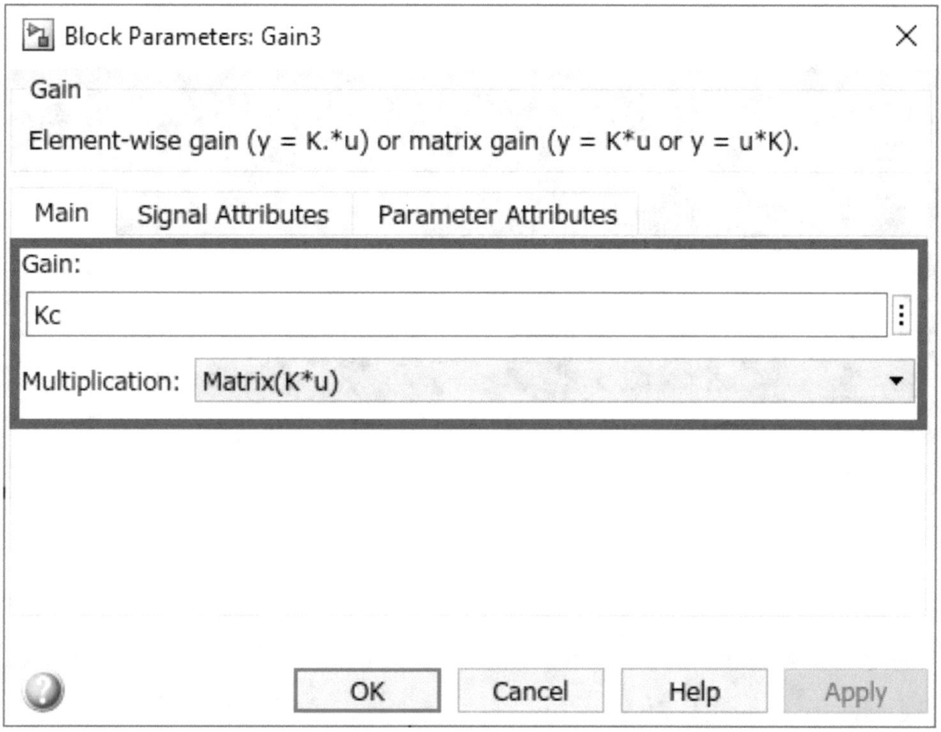

Figure 3.27: The setting of the block shown with number 3 in Fig. 3.24.

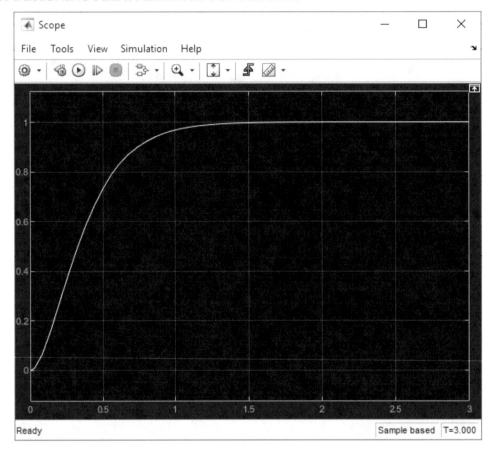

Figure 3.28: Simulation result.

The state-space model of system is:

$$
\frac{d}{dt}\begin{bmatrix} \theta \\ \dot{\theta} \\ i \end{bmatrix} = \begin{bmatrix} 0 & 1 & 0 \\ 0 & -\dfrac{b}{J} & \dfrac{K}{J} \\ 0 & -\dfrac{K}{J} & -\dfrac{R}{L} \end{bmatrix} \begin{bmatrix} \theta \\ \dot{\theta} \\ i \end{bmatrix} + \begin{bmatrix} 0 \\ 0 \\ \dfrac{1}{L} \end{bmatrix} V
$$

$$
y = \begin{bmatrix} 1 & 0 & 0 \end{bmatrix} \begin{bmatrix} \theta \\ \dot{\theta} \\ i \end{bmatrix},
$$

(3.10)

where J is the moment of inertia of the rotor and equals to 3.2284 μkg·m^2, b is the motor viscous friction constant and equals to 3.5077 μN·m·s, K is the motor torque constant and equals to 0.0274 N·m/A, R is the armature resistance and equals to 4 Ω, and L is the armature inductance and equals to 2.75 μH.

We want to be able to position the motor very precisely, thus the steady-state error of the motor position should be zero when given a commanded position. The other performance requirement is that the motor reaches its final position very quickly without excessive overshoot. In this case, we want the system to have a settling time of 40 ms and an overshoot smaller than 16%.

3.3.1 DESIGNING THE FULL-STATE FEEDBACK CONTROLLER

The following code defines the state-space model of system and calculate the required gains for placing the closed-loop poles at -200, $-100+100i$, and $-100-100i$. By ignoring the effect of the first pole (since it is faster than the other two poles), the dominant poles correspond to a second-order system with $\zeta = 0.5$ corresponding to 16% overshoot and $\sigma = 100$ which corresponds to a settling time of 0.040 s. The required gain values for full state feedback is shown in Fig. 3.29.

```
J = 3.2284E-6;
b = 3.5077E-6;
K = 0.0274;
R = 4;
L = 2.75E-6;

A = [0 1 0
     0 -b/J K/J
     0 -K/L -R/L];
B = [0 ; 0 ; 1/L];
C = [1  0  0];
D = 0;
motor_ss = ss(A,B,C,D);

p1 = -100+100i;
p2 = -100-100i;
p3 = -200;
Kc = place(A,B,[p1, p2, p3])
```

With the aid of command shown in Fig. 3.30, you can ensure that the gains given in Fig. 3.29 place the poles at the desired locations.

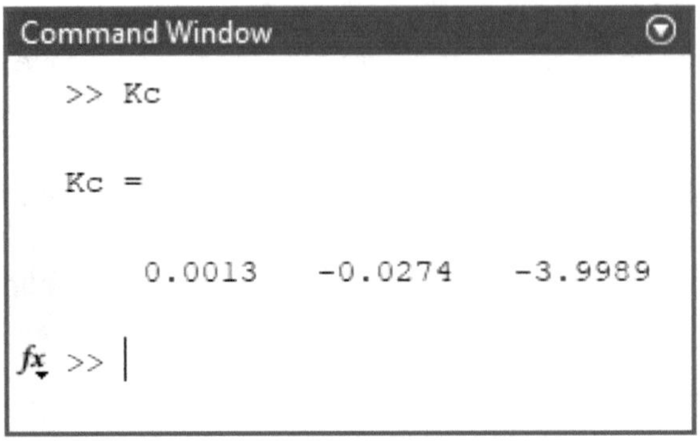

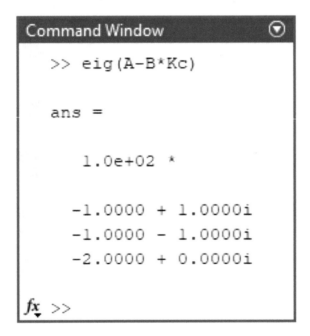

Figure 3.29: Required gains.

Figure 3.30: The calculated gains place the poles at –100+100i, –100–100i, and –200.

We use the Simulink to simulate the system. The Simulink model is shown in Fig. 3.31. The setting of blocks are shown in Figs. 3.32–3.37. The simulation result is shown in Fig. 3.38.

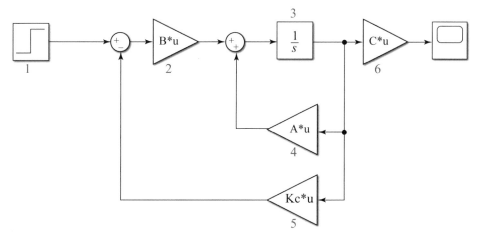

Figure 3.31: **Simulink diagram of system.**

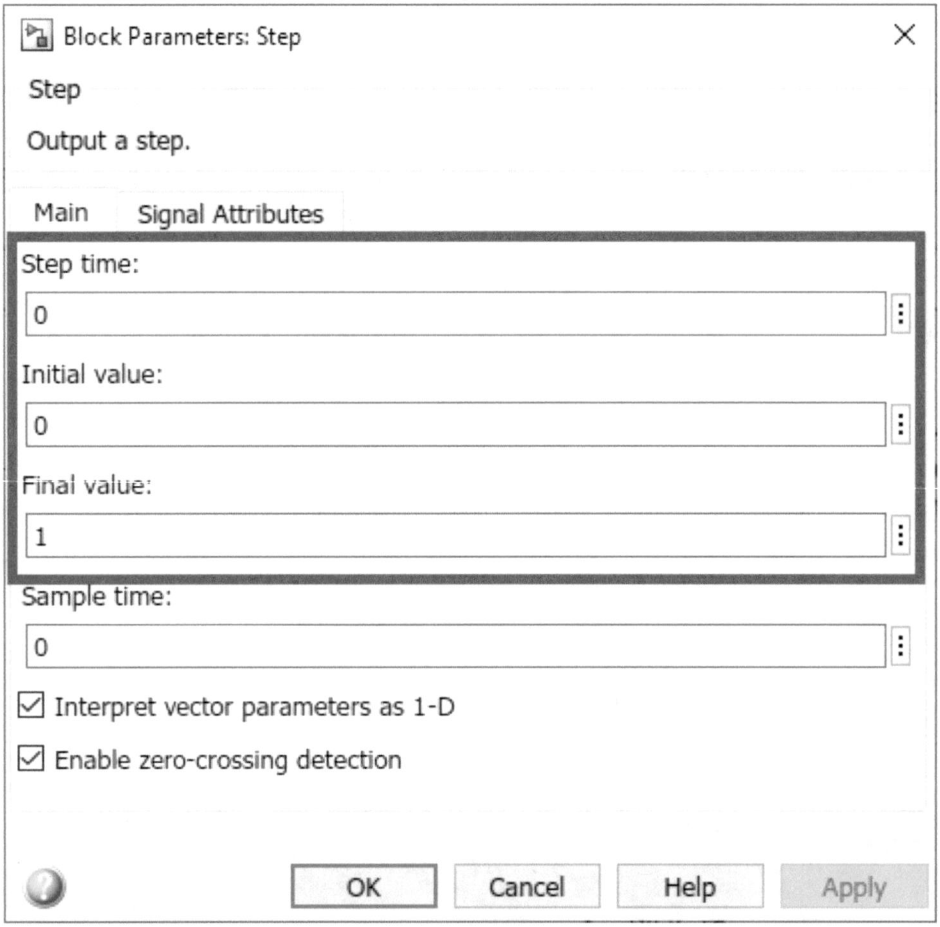

Figure 3.32: The setting of the block shown with number 1 in Fig. 3.31.

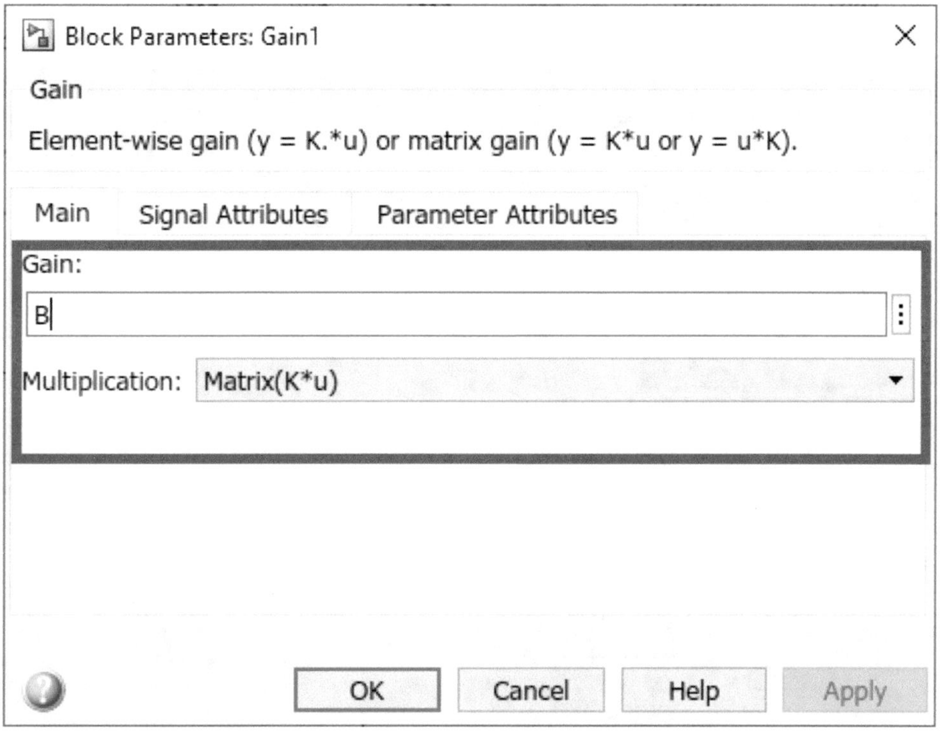

Figure 3.33: The setting of the block shown with number 2 in Fig. 3.31.

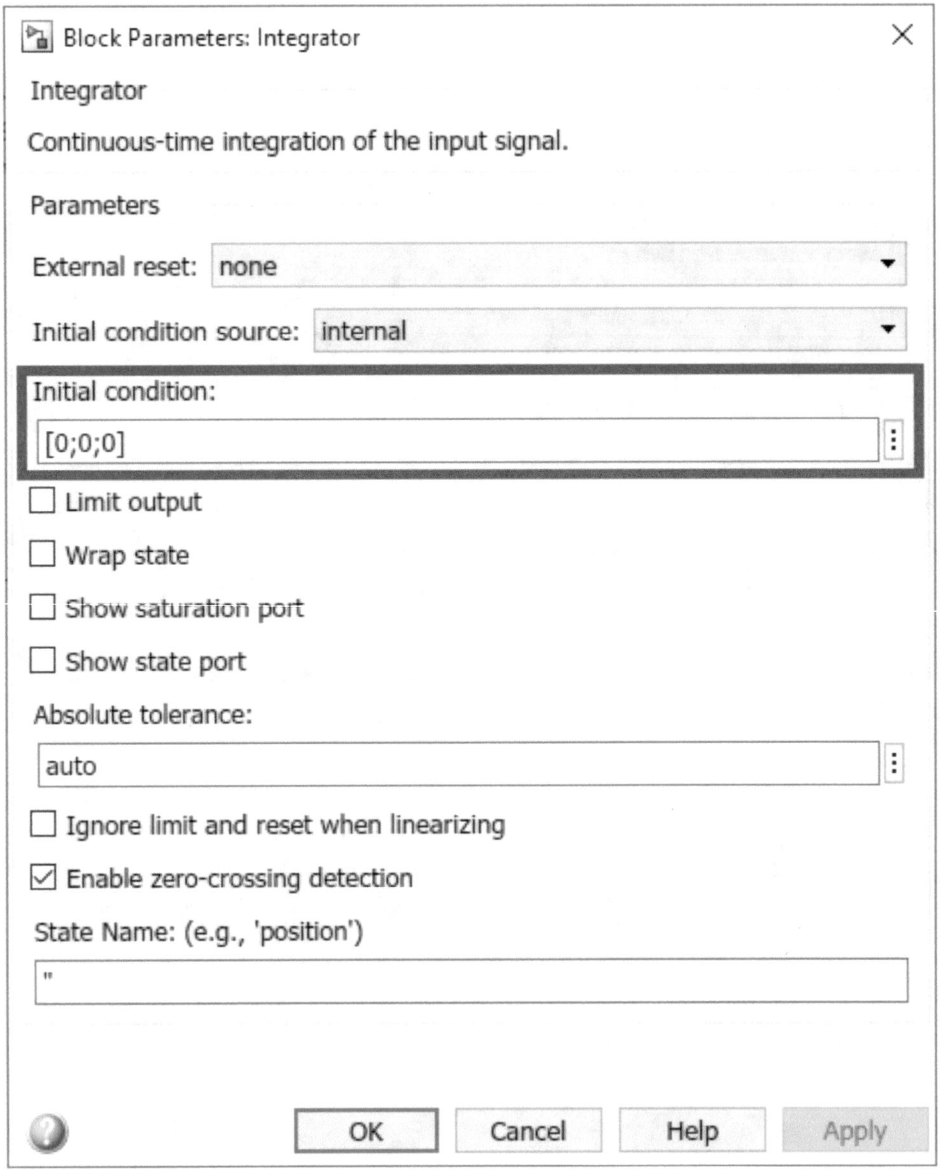

Figure 3.34: The setting of the block shown with number 3 in Fig. 3.31.

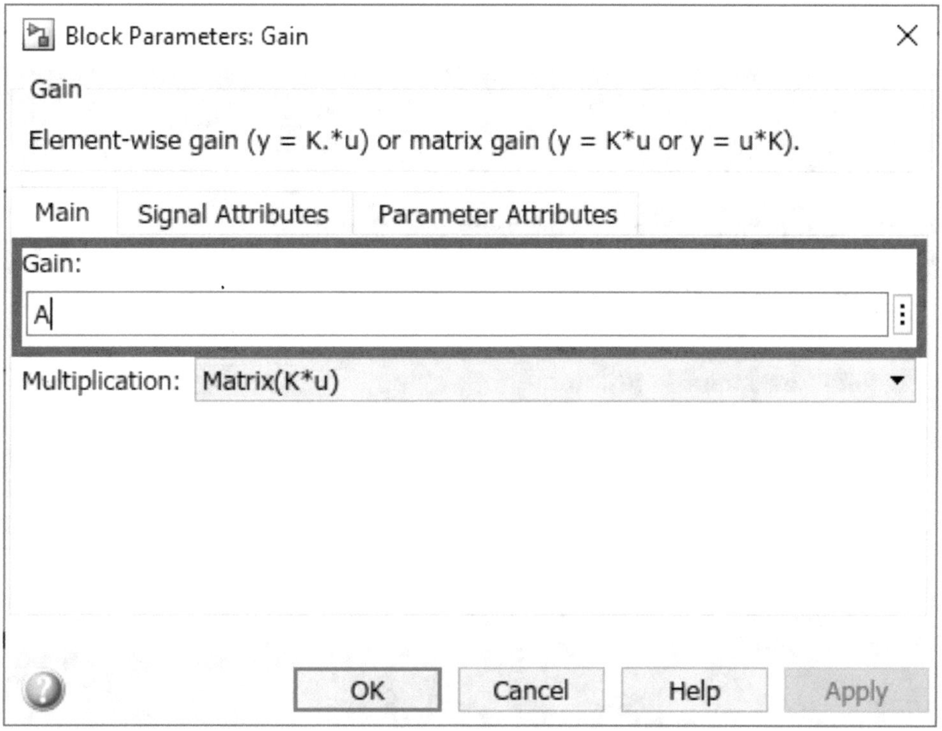

Figure 3.35: The setting of the block shown with number 4 in Fig. 3.31.

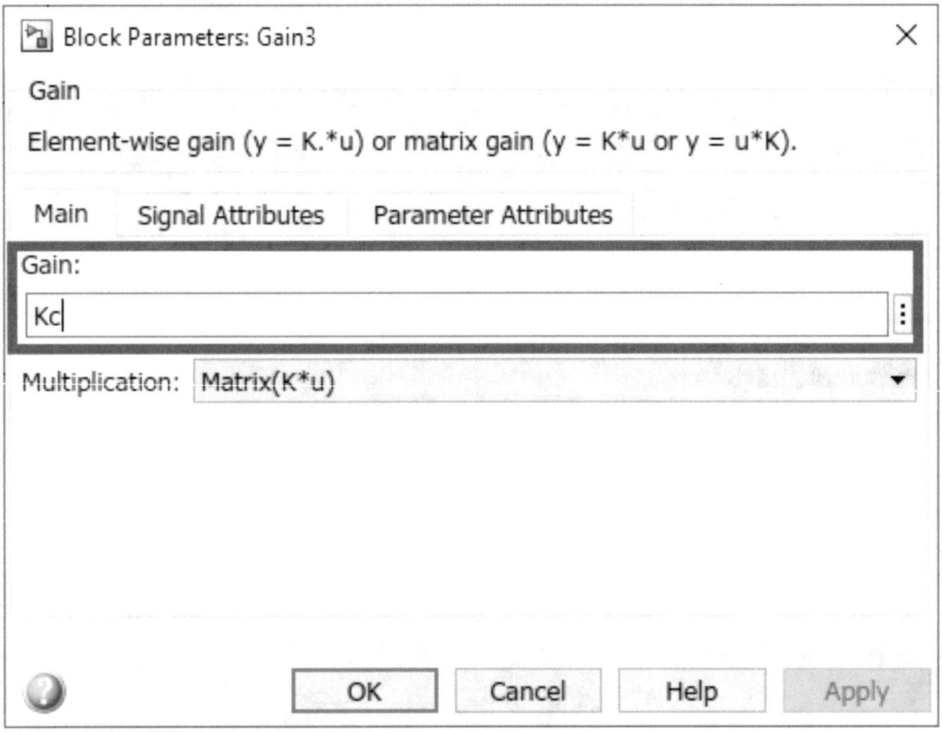

Figure 3.36: The setting of the block shown with number 5 in Fig. 3.31.

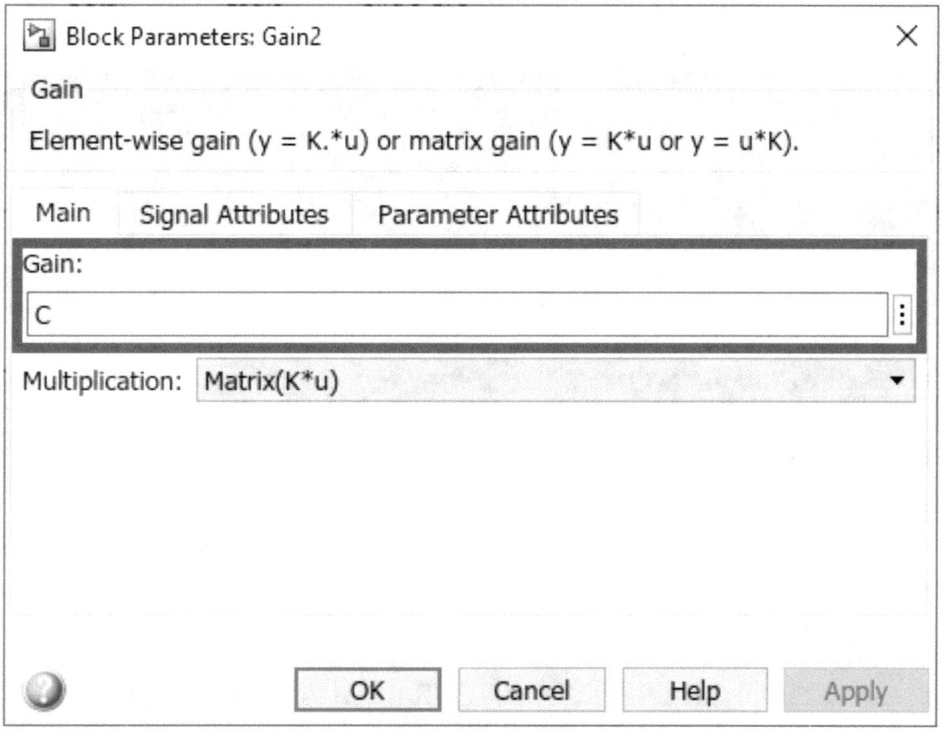

Figure 3.37: The setting of the block shown with number 6 in Fig. 3.31.

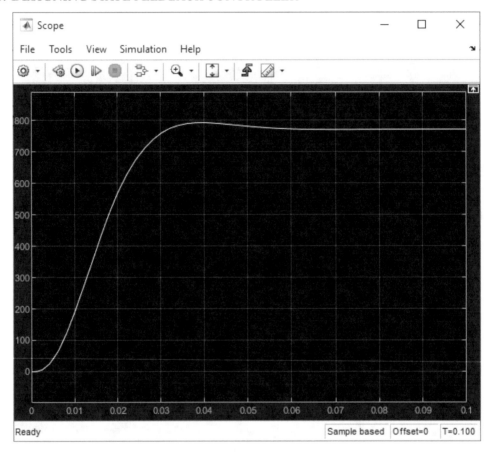

Figure 3.38: Simulation result.

3.3.2   ADDITION OF AN INTEGRATOR BLOCK FOR SOLVING THE STEADY-STATE ERROR PROBLEM

As seen in Fig. 3.38, the closed-loop system has considerable steady-state error. In order to solve the steady-state problem of closed-loop system, we add an integrator to the system as shown in Fig. 3.39. We know that if we put an extra integrator in series with the plant, it can remove the steady-state error due to a step reference. Addition of an integrator to the system, increase the number of states by one. The gains K_c and K_i in Fig. 3.39 are unknown and need to be found.

   Addition of integrator changes the control structure. We can model the addition of this integrator by augmenting our state equations with an extra state for the integral of the error which we will identify with the variable ω. This adds an extra state equation, where the derivative of this state is then just the error, $e = y - r$ where $y = \theta$. This equation will be placed at the bottom of our matrices. The reference r, therefore, now appears as an additional input to our

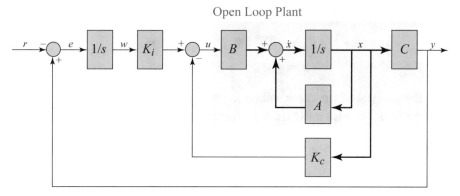

Figure 3.39: Addition of an integrator for solving the problem of steady-state error.

system. The output of the system remains the same. So,

$$
\frac{d}{dt}\begin{bmatrix} \theta \\ \dot{\theta} \\ i \\ w \end{bmatrix} = \begin{bmatrix} 0 & 1 & 0 & 0 \\ 0 & -b/J & K/J & 0 \\ 0 & -K/L & -R/L & 0 \\ 1 & 0 & 0 & 0 \end{bmatrix} \begin{bmatrix} \theta \\ \dot{\theta} \\ i \\ w \end{bmatrix} + \begin{bmatrix} 0 \\ 0 \\ 1/L \\ 0 \end{bmatrix} V + \begin{bmatrix} 0 \\ 0 \\ 0 \\ -1 \end{bmatrix} r
$$

$$
y = \begin{bmatrix} 1 & 0 & 0 & 0 \end{bmatrix} \begin{bmatrix} \theta \\ \dot{\theta} \\ i \\ w \end{bmatrix}.
$$

(3.11)

These equations represent the dynamics of the system before the loop is closed. We will refer to the system matrices in this equation that are augmented with the additional integrator state as A_a, B_a, C_a, and D_a. The vector multiplying the reference input r will be referred to as B_r. We will refer to the state vector of the augmented system as x_a. Note that the reference, r, does not affect the states (except the integrator state) or the output of the plant. This is expected since there is no path from the reference to the plant input, u, without implementing the state-feedback gain matrix K_c.

In order to find the closed-loop equations, we have to look at how the input, u, affects the plant. In this case, it affects the system in exactly the same manner as in the unaugmented equations except now $u = -K_c x - K_i w$. We can also rewrite this in terms of our augmented state as $u = -K_a x_a$ where $K_a = [K_c \ K_i]$. Substituting this u into the equations above provides the following closed-loop equations:

$$\dot{x}_a = (A_a - B_a K_a) x_a + B_r r$$
$$y = C_a x_a.$$

(3.12)

In the above, the integral of the error will be fed back, and will result in the steady-state error being reduced to zero. Now we must redesign our controller to account for the augmented state vector. Since we need to place each pole of the system, we will place the pole associated with the additional integrator state at −300, which will be faster than the other poles.

The following code, calculates the required gains (Fig. 3.40):

```
J = 3.2284E-6;
b = 3.5077E-6;
K = 0.0274;
R = 4;
L = 2.75E-6;

Aa = [0 1 0 0
        0 -b/J K/J 0
        0 -K/L -R/L 0
        1 0 0 0];
Ba = [0 ; 0 ; 1/L ; 0 ];
Br = [0 ; 0 ; 0; -1];
Ca = [1  0  0  0];
Da = [0];

p1 = -100+100i;
p2 = -100-100i;
p3 = -200;
p4 = -300;

Ka = place(Aa,Ba,[p1,p2,p3,p4]);
```

We use the Simulink diagram shown in Fig. 3.41 in order to test the closed-loop system. The setting of blocks are shown in Figs. 3.42–3.49.

After running the simulation, the result shown in Fig. 3.50 is obtained. As seen, the closed-loop system has zero steady-state error.

Instead of using the Simulink diagram shown in Fig. 3.41, you can use the following code to obtain the step response of closed-loop system (see Fig. 3.51).

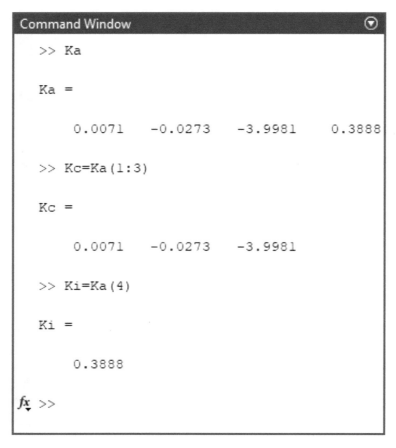

Figure 3.40: Calculated required gains.

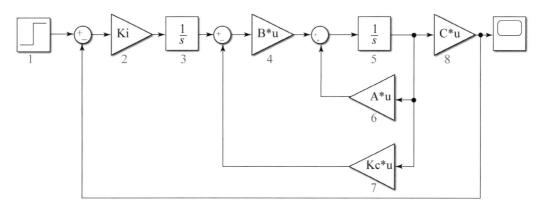

Figure 3.41: Simulink diagram of system.

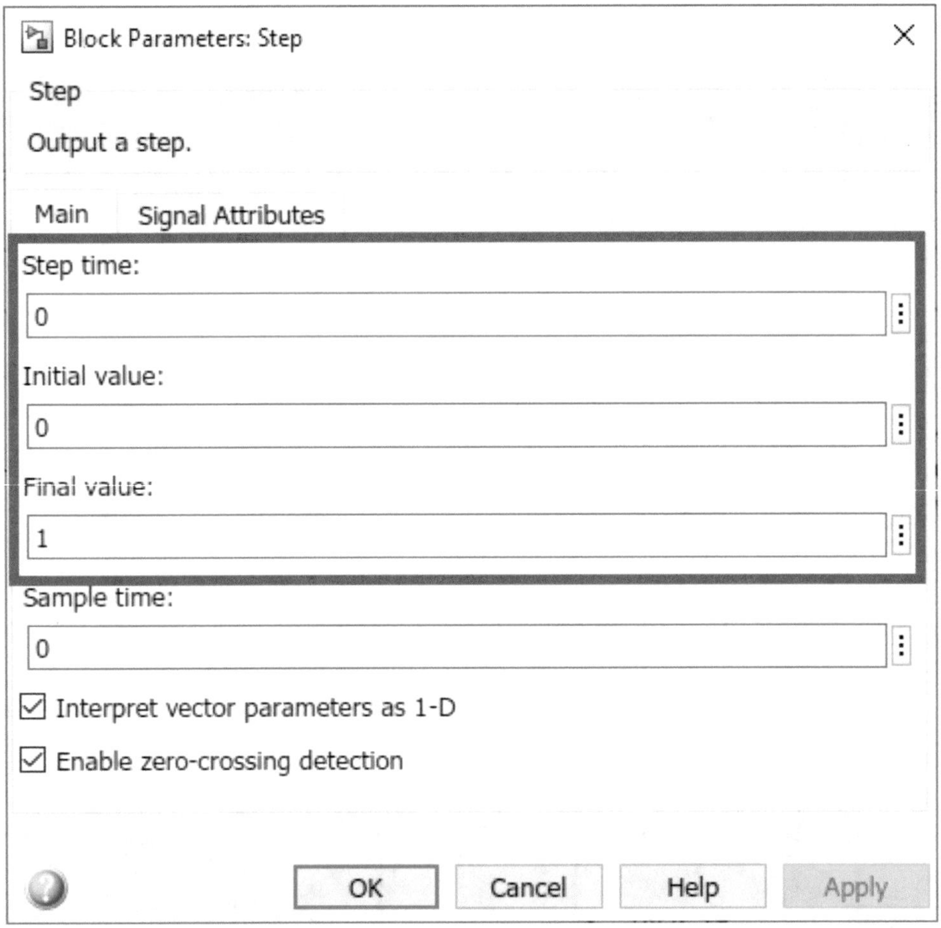

Figure 3.42: The setting of the block shown with number 1 in Fig. 3.41.

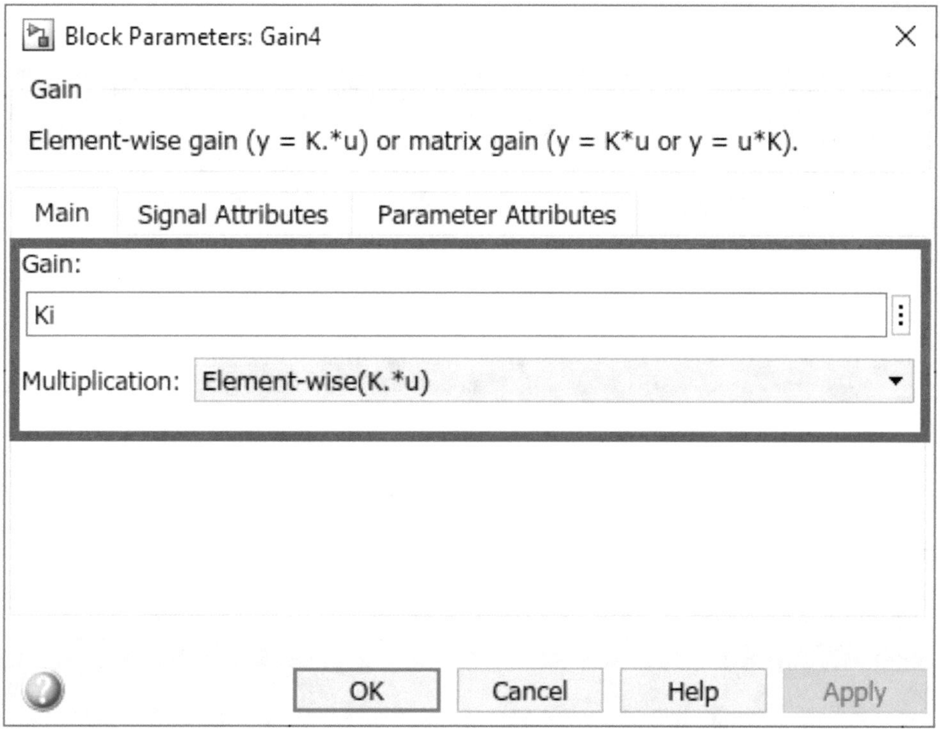

Figure 3.43: The setting of the block shown with number 2 in Fig. 3.41.

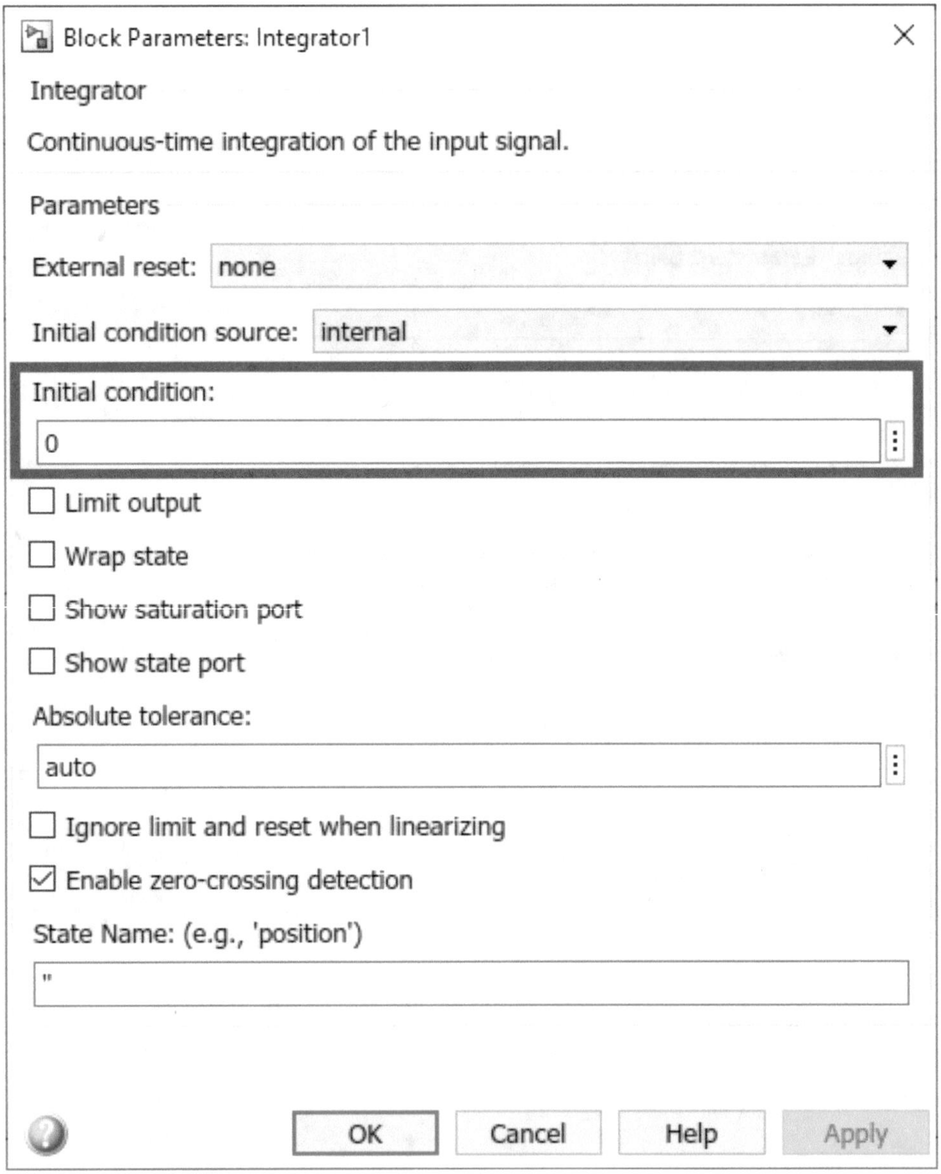

Figure 3.44: The setting of the block shown with number 3 in Fig. 3.41.

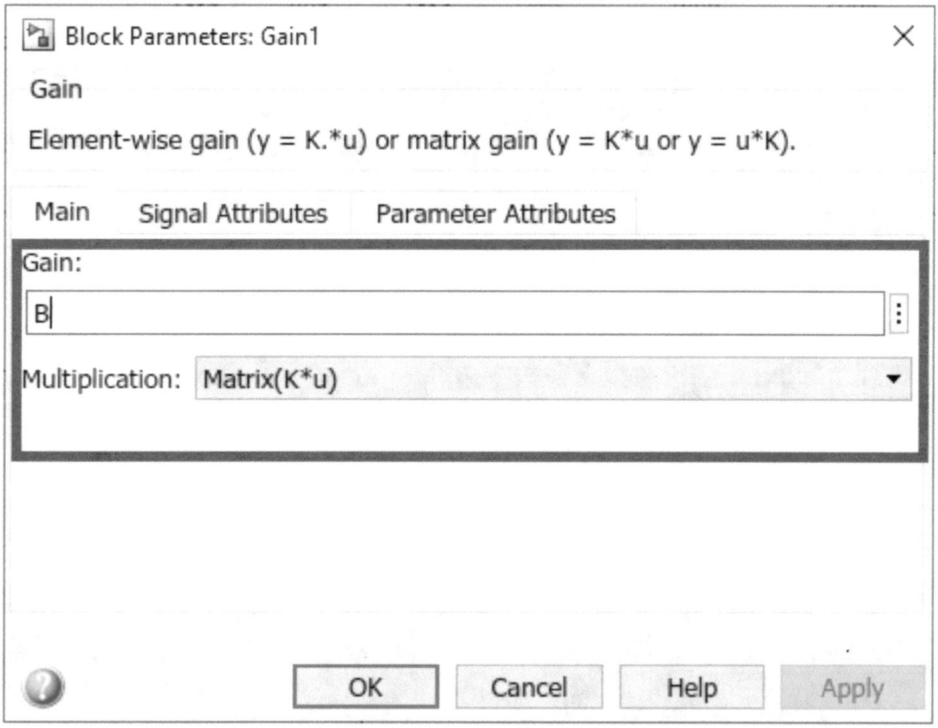

Figure 3.45: The setting of the block shown with number 4 in Fig. 3.41.

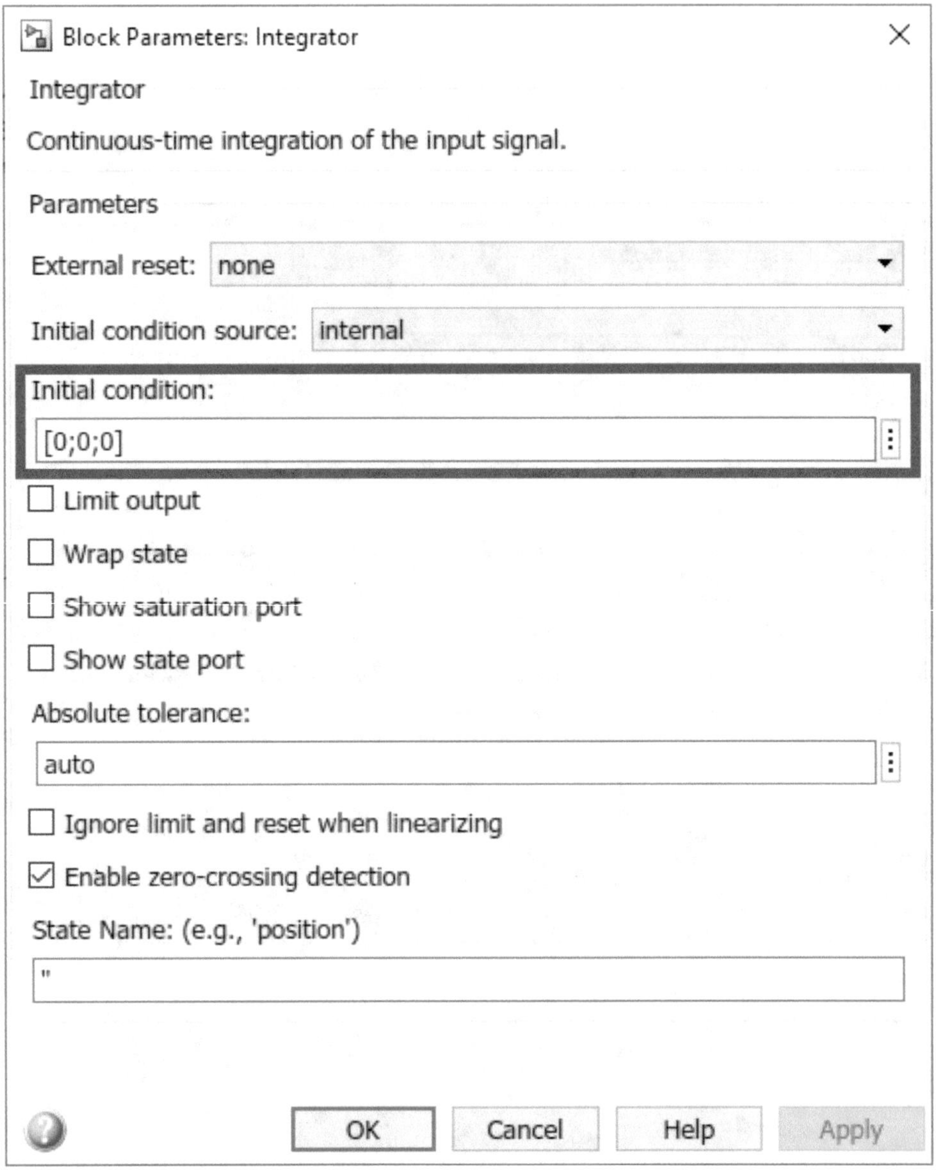

Figure 3.46: The setting of the block shown with number 5 in Fig. 3.41.

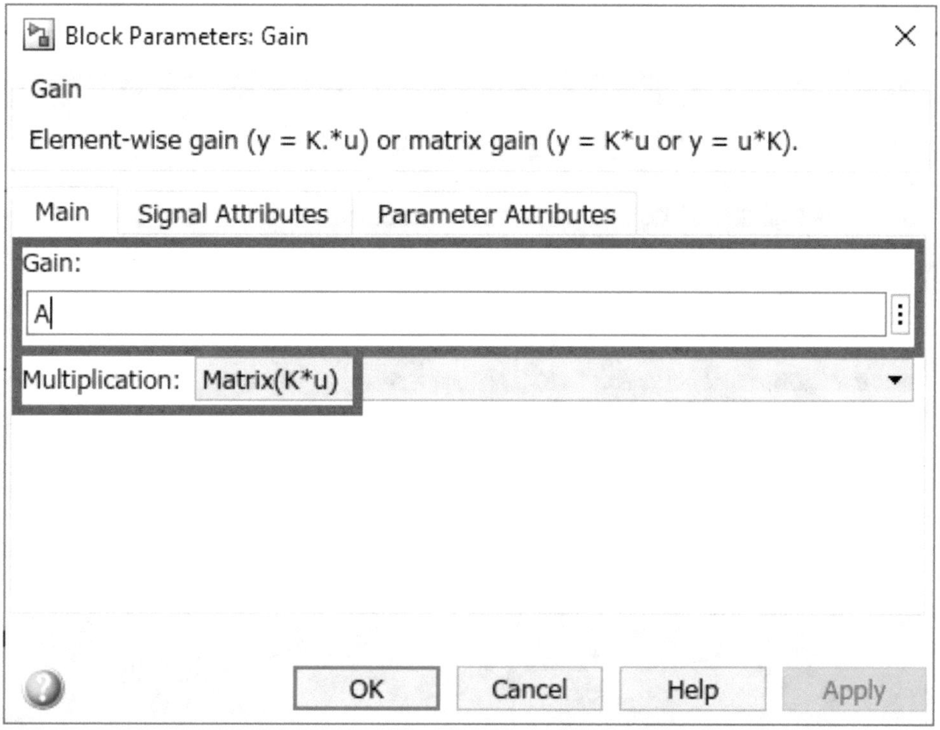

Figure 3.47: The setting of the block shown with number 6 in Fig. 3.41.

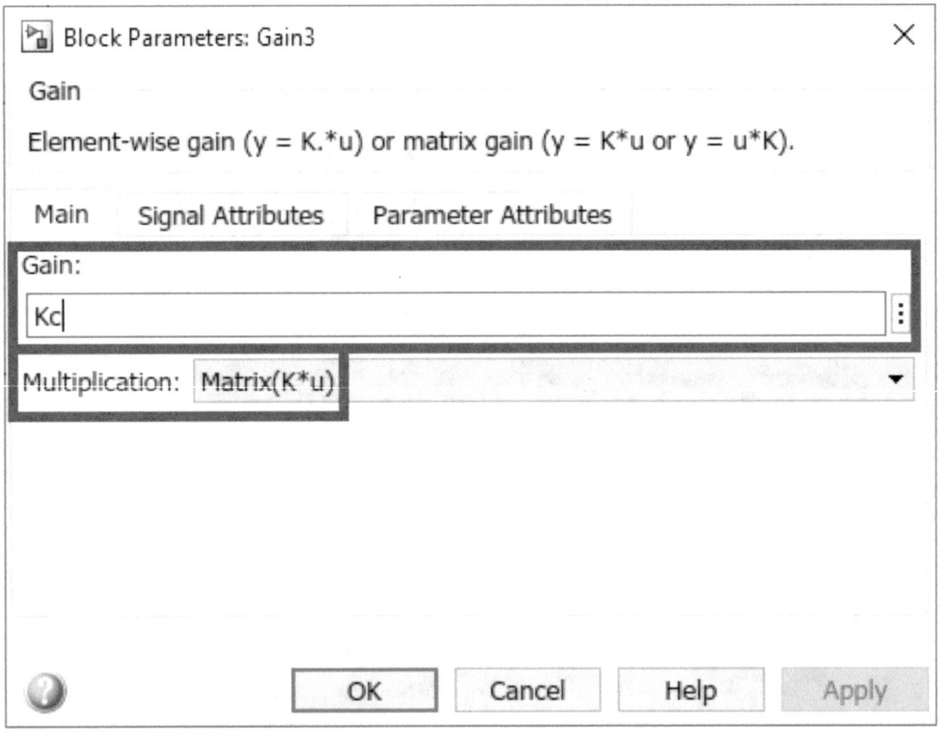

Figure 3.48: The setting of the block shown with number 7 in Fig. 3.41.

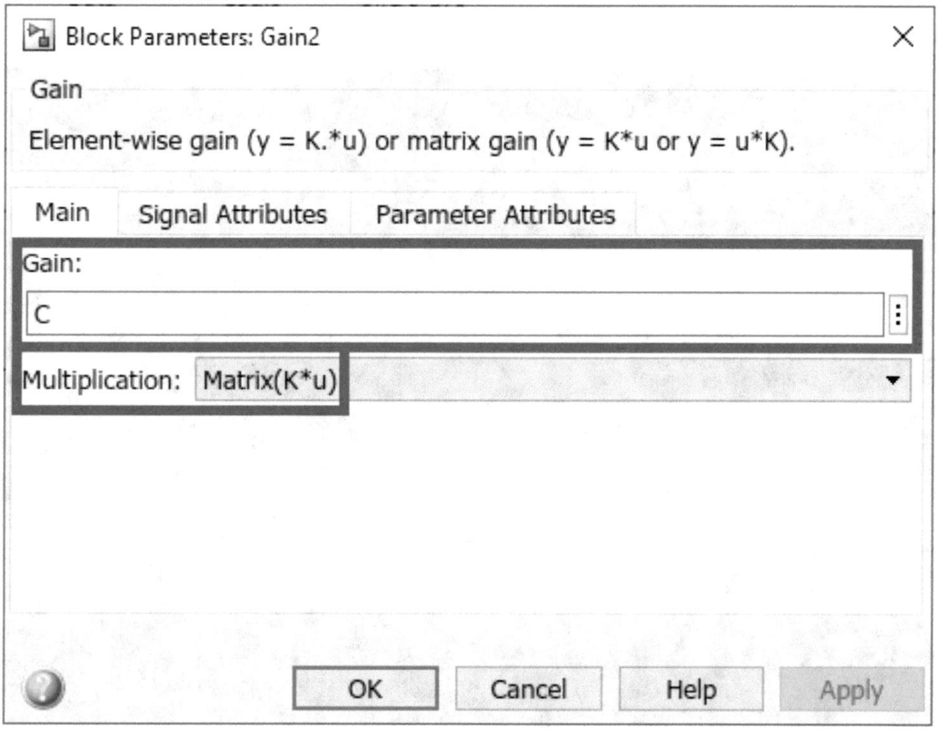

Figure 3.49: The setting of the block shown with number 8 in Fig. 3.41.

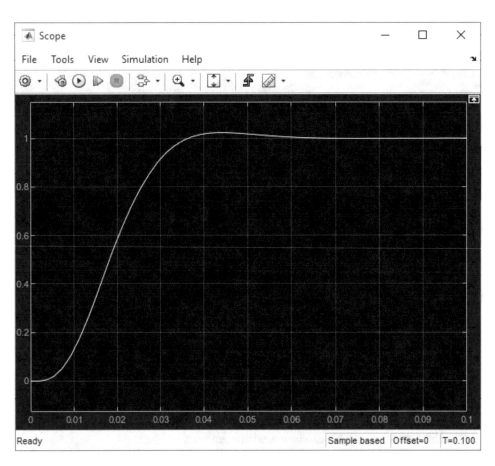

Figure 3.50: Simulation result.

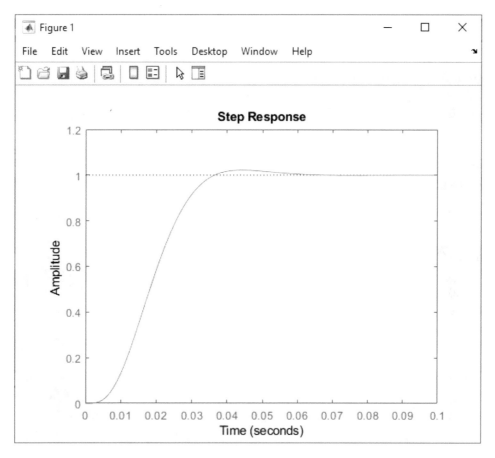

Figure 3.51: Result of running the code.

```
J = 3.2284E-6;
b = 3.5077E-6;
K = 0.0274;
R = 4;
L = 2.75E-6;

Aa = [0 1 0 0
      0 -b/J K/J 0
      0 -K/L -R/L 0
      1 0 0 0];
Ba = [0 ; 0 ; 1/L ; 0 ];
Br = [0 ; 0 ; 0; -1];
Ca = [1  0  0  0];
Da = [0];

p1 = -100+100i;
p2 = -100-100i;
p3 = -200;
p4 = -300;

Ka = place(Aa,Ba,[p1,p2,p3,p4]);

t = 0:0.001:.1;
sys_cl = ss(Aa-Ba*Ka,Br,Ca,Da);
step(sys_cl,t)
```

3.4   EXAMPLE 3: MAGNETIC LEVITATION SYSTEM

Magnetic levitation or magnetic suspension is a method by which an object is suspended with no support other than magnetic fields.

3.4.1   MODELING THE SYSTEM

A magnetic levitation system is shown in Fig. 3.52. The current through the coils induces a magnetic force which can balance the force of gravity and cause the ball (which is made of a magnetic material) to be suspended in mid-air.

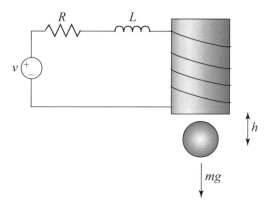

Figure 3.52: Magnetic levitation system.

The equations for the system are given by:

$$m\frac{d^2h}{dt^2} = mg - \frac{Ki^2}{mg}$$

$$V = L\frac{di}{dt} + iR,$$

(3.13)

where h is the vertical position of the ball, i is the current through the electromagnet, V is the applied voltage, m is the mass of the ball, g is the acceleration due to gravity, L is the inductance, R is the resistance, and K is a coefficient that determines the magnetic force exerted on the ball.

For simplicity, we will choose values $m = 0.05$ kg, $K = 0.0001$, $L = 0.01$ H, $R = 1$ Ohm, and $g = 9.81$ m/s^2. The system is at equilibrium (the ball is suspended in mid-air) whenever $h = \frac{Ki^2}{mg}$.

We linearize the equations about the point $h = 0.01$ m (where the nominal current is about $i = \sqrt{\frac{h \times m \times g}{K}} = \sqrt{\frac{0.01 \times 0.05 \times 9.8}{0.0001}} = 7$ Amps) and obtain the linear state-space equations:

$$\frac{dx}{dt} = Ax + Bu$$

$$y = Cx + Du,$$

(3.14)

where

$$x = \begin{bmatrix} \Delta h \\ \Delta \dot{h} \\ \Delta i \end{bmatrix}$$

(3.15)

is the set of state variables for the system (a 3×1 vector), u is the deviation of the input voltage from its equilibrium value (ΔV), and y (the output) is the deviation of the height of the ball from its equilibrium position (Δh).

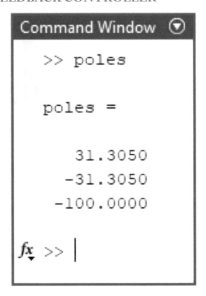

Figure 3.53: Open loop poles of the magnetic levitation system.

The following commands define the system matrices and calculate the eigen values of matrix A (which are the system poles):

```
A = [ 0    1    0
      980  0    -2.8
      0    0    -100 ];

B = [ 0
      0
      100 ];

C = [1 0 0 ];

poles = eig(A);
```

After running the code, the result shown in Fig. 3.53 is obtained.

Values shown in Fig. 3.53 show that the open-loop system is unstable (since it has a pole in the RHP). The following code shows what happens to this unstable system when there is a non-zero initial condition. After running the code, the result shown in Fig. 3.54 will be obtained.

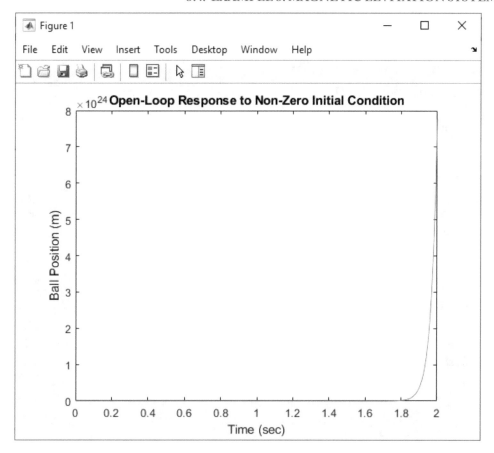

Figure 3.54: Step response of open-loop magnetic levitation system.

```
A = [ 0     1    0
      980   0   -2.8
      0     0   -100 ];

B = [ 0
      0
      100 ];

C = [1 0 0];

t = 0:0.01:2;
u = zeros(size(t));
```

```
x0 = [0.01 0 0];

sys = ss(A,B,C,0);

[y,t,x] = lsim(sys,u,t,x0);
plot(t,y)
title('Open-Loop Response to Non-Zero
        Initial Condition')
xlabel('Time (sec)')
ylabel('Ball Position (m)')
```

According to Fig. 3.54, it looks like the distance between the ball and the electromagnet will go to infinity, but probably the ball hits the table or the floor first (and also, probably goes out of the range where our linearization is valid).

3.4.2   DESIGNING THE FULL-STATE FEEDBACK CONTROLLER

Let's build a full-state feedback controller for this system using the pole placement approach. Assume that we decide to place the poles ate $-20 \pm 20i$ and -100. Using the following code, calculate the required gains. In general, the farther you move the poles to the left, the more control effort is required:

```
A = [ 0    1   0
      980  0  -2.8
      0    0  -100 ];

B = [ 0
      0
      100 ];

C = [1  0  0];

p1 = -20 + 20i;
p2 = -20 - 20i;
p3 = -100;

K = place(A,B,[p1 p2 p3]);
```

After running the code, the gains shown in Fig. 3.55 are obtained.

As shown in Fig. 3.56, you can use the eig(A-B*K) command to ensure that the given gains put the closed-loop poles at the desired location.

```
Command Window                          ⊙
    >> K

    K =

        -775.7143   -20.6429    0.4000

fx >> |
```

Figure 3.55: Required feedback gains for full-state feedback.

```
Command Window                          ⊙
    >> eig(A-B*K)

    ans =

        1.0e+02 *

        -1.0000 + 0.0000i
        -0.2000 + 0.2000i
        -0.2000 - 0.2000i

fx >> |
```

Figure 3.56: Eigen values (poles) of the closed-loop system.

The Simulink model shown in Fig. 3.57 is used for simulating the system. The simulation result is shown in Fig. 3.58.

3.4.3   ADDITION OF A GAIN BLOCK FOR SOLVING THE STEADY-STATE ERROR PROBLEM

According to Fig. 3.58, the system does not track the step reference well at all; not only is the magnitude not one, but it is negative instead of positive! Recall in the schematic above (Fig. 3.57), that we don't compare the output to the reference; instead we measure all the states, multiply by the gain vector K, and then subtract this result from the reference. There is no reason to expect that Kx will be equal to the desired output.

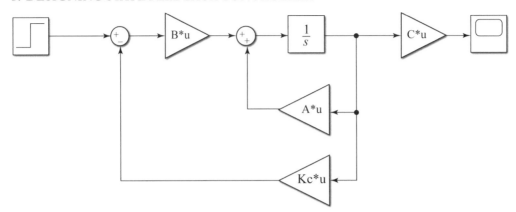

Figure 3.57: Simulink model of magnetic levitation system with full-state feedback.

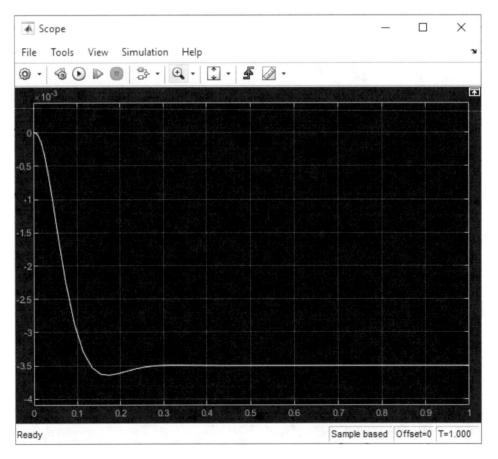

Figure 3.58: Step response of the model shown in Fig. 3.57. The steady-state value is −0.0035.

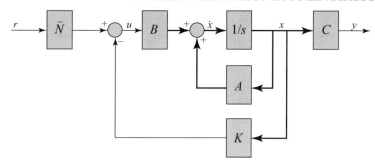

Figure 3.59: Addition of block $\overline{N}$ to the full state feedback system.

Figure 3.60: Calculation of the gain $\overline{N}$.

To eliminate this problem, we can scale the reference input to make it equal to Kx in steady-state. The scale factor, $\overline{N}$, is shown in the schematic; see Fig. 3.59.

Value of $\overline{N}$ could be calculated as shown in Fig. 3.60. Figure 3.61 shows the updated Simulink model. After simulating this diagram, the result shown in Fig. 3.62 is obtained. As seen, the output tracks the step response.

3.4.4   DESIGNING THE OBSERVER

When we cannot measure all state variables x (often the case in practice), we can build an observer to estimate them, while measuring only the output $y = Cx$. For the magnetic ball example, we will add three new, estimated state variables ($\hat{x}$) to the system. The schematic is as shown in Fig. 3.63.

The observer is basically a copy of the plant; it has the same input and almost the same differential equation. An extra term compares the actual measured output y to the estimated

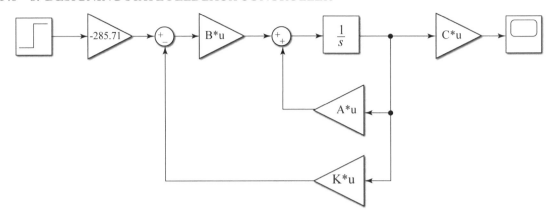

Figure 3.61: Addition of gain $\overline{N}$ to the Simulink model.

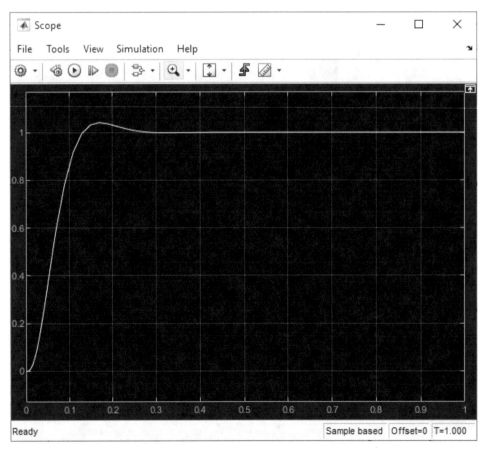

Figure 3.62: Result of simulating the model shown in Fig. 3.61.

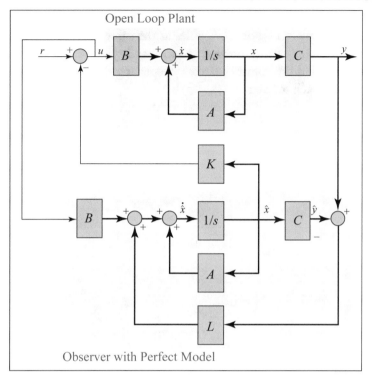

Figure 3.63: Observer-based control system.

output $\hat{y} = C\hat{x}$; this will help to correct the estimated state $\hat{x}$ and cause it to approach the values of the actual state x (if the measurement has minimal error).

According to Fig. 3.63,

$$\dot{\hat{x}} = A\hat{x} + Bu + L(y - \hat{y})$$
$$\hat{y} = C\hat{x}$$

(3.16)

and

$$\begin{cases} \dot{x} = Ax + Bu \\ y = Cx. \end{cases}$$

(3.17)

So,

$$\dot{e} = \dot{x} - \dot{\hat{x}} = (A - LC)e,$$

(3.18)

which means the error dynamics of the observer are given by the poles of $A - LC$.

In order to design the observer, we need to choose the observer gain L first. Since we want the dynamics of the observer to be much faster than the system itself, we need to place the poles at least five times farther to the left than the dominant poles of the system.

If we want to use `place` command, we need to put the three observer poles at different locations. If you want to place two or more poles at the same position, you can use a function called `acker`. Because of the duality between controllability and observability, we can use the same technique used to find the control matrix by replacing the matrix *B* by the matrix *C* and taking the transposes of each matrix:

```
A = [ 0    1   0
      980  0  -2.8
      0    0  -100 ];

B = [ 0
      0
      100 ];

C = [ 1  0  0 ];

op1 = -100;
op2 = -101;
op3 = -102;

L = place(A',C',[op1 op2 op3])';
```

Value of *L* is shown in Fig. 3.64.

The simulation diagram of the observer-based control system is shown in Fig. 3.65. Note that we used the estimated states for feedback, i.e., $u = -K\hat{x}$. In Fig. 3.61, we used the real states for making the control input, i.e., $u = -Kx$.

Let us simulate the system and compare the output of designed observer with the real system states. Assume that the initial states of the plant (upper integrator) are [0.01 0.5 -5] (Fig. 3.66).

The initial state of observer may not be the same as plant. Let us simply assume that the initial states of observer system (lower integrator) is [0 0 0] (Fig. 3.67).

The simulation results are shown in Figs. 3.68–3.71. According to Figs. 3.68–3.70, the designed observer converges to real values of states. Figure 3.71 shows the output of system.

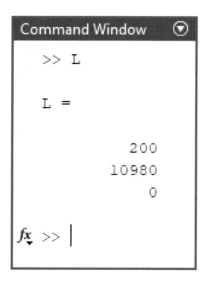

Figure 3.64: Values of vector L.

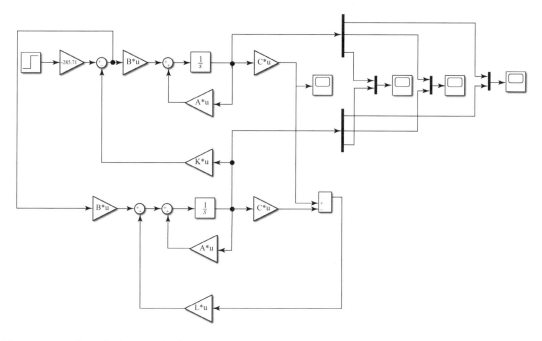

Figure 3.65: Simulink model of the closed-loop magnetic levitation system with observer-based controller.

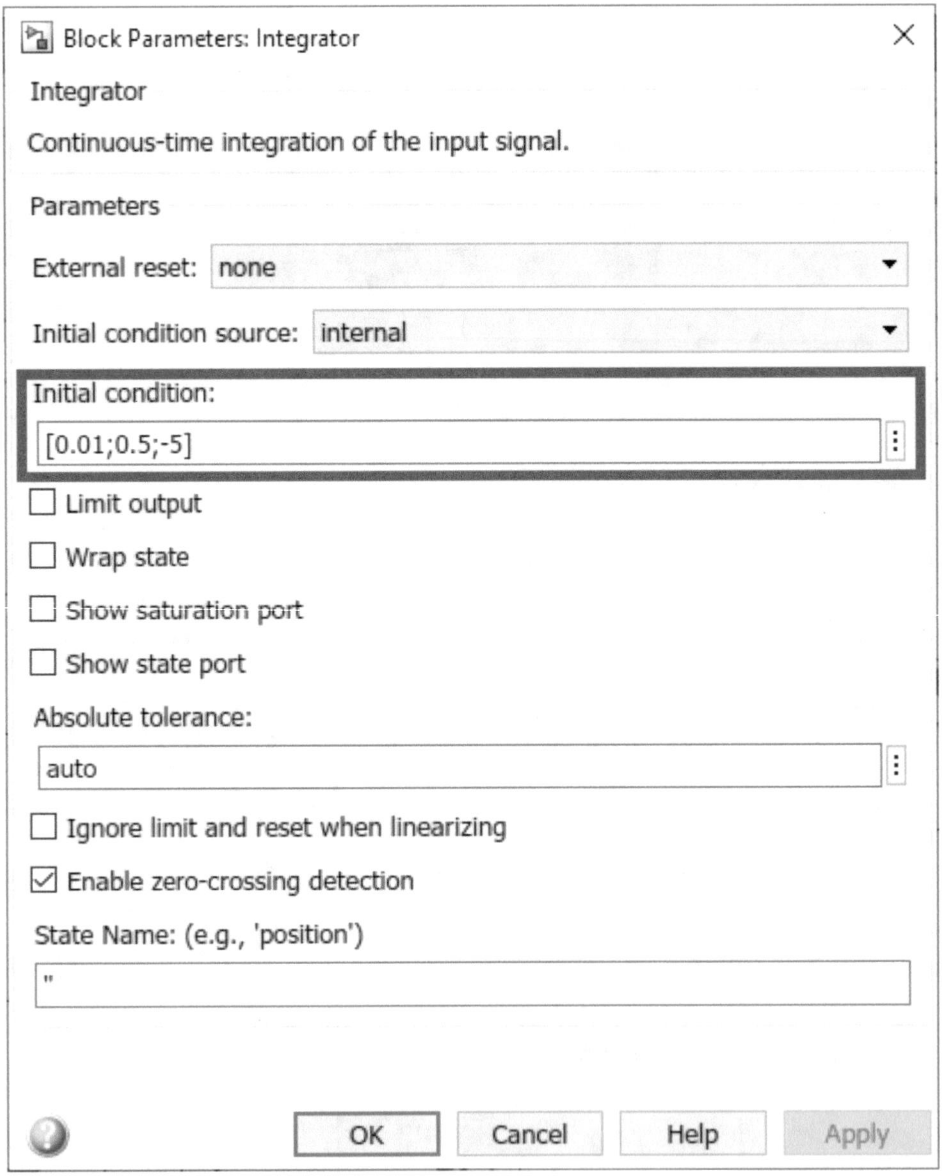

Figure 3.66: Settings of the upper integrator (plant) in Fig. 3.65.

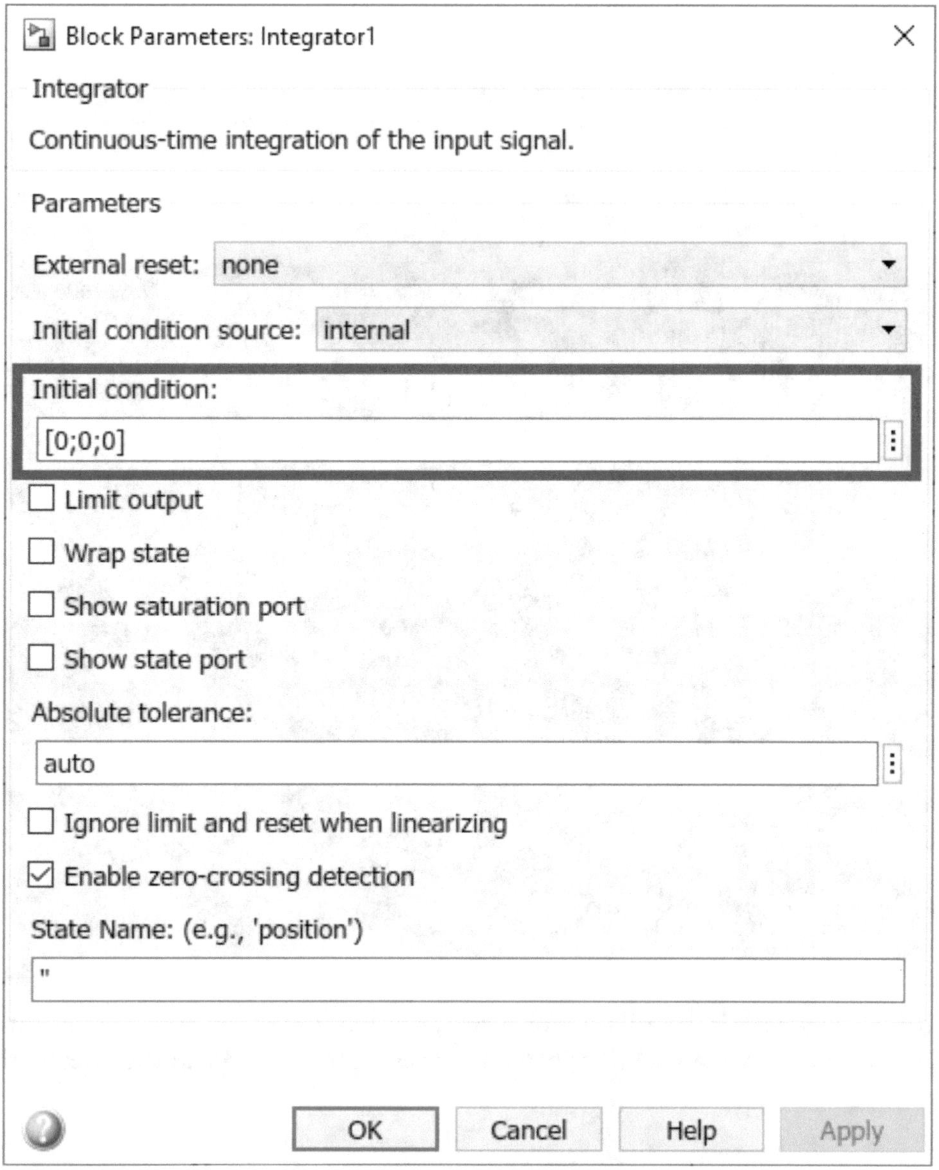

Figure 3.67: Settings of the lower integrator (observer) in Fig. 3.65.

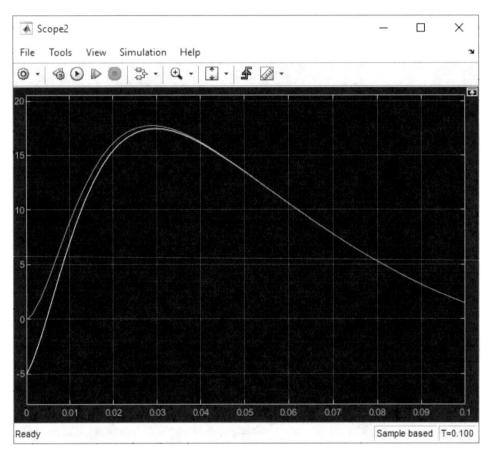

Figure 3.68: Yellow trace is the real value of state x_1 and blue trace shows the output of observer (estimated value $\hat{x}_1$). For $t > 0.04$ s, the two traces overlap. So, the estimation converges the real value.

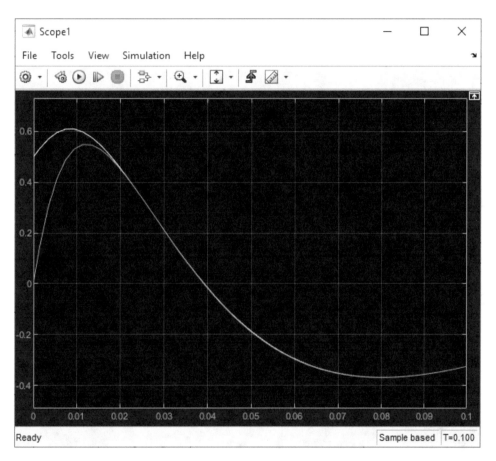

Figure 3.69: Yellow trace is the real value of state x_2 and blue trace shows the output of observer (estimated value $\hat{x}_2$). For $t > 0.02$ s, the two traces overlap. So, the estimation converges the real value.

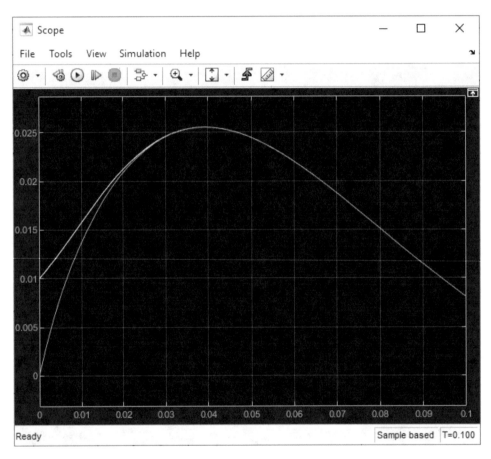

Figure 3.70: Yellow trace is the real value of state x_3 and blue trace shows the output of observer (estimated value $\hat{x}_3$). For $t > 0.03$ s, the two traces overlap. So, the estimation converges the real value.

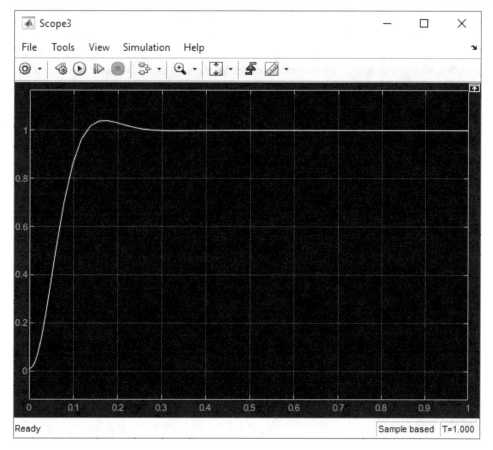

Figure 3.71: **Step response of a closed-loop control system shown in Fig.** 3.65.

3.5   REFERENCES

[1] http://ctms.engin.umich.edu/CTMS/index.php?aux=Home 89, 155

Website [1] is a collection of examples developed by Prof. Bill Messner (Department of Mechanical Engineering at Tufts University) and Prof. Dawn Tilbury (Department of Mechanical Engineering and Applied Mechanics at the University of Michigan).

Most of the original development work was done by undergraduate students Luis Oms (CMU), Joshua Pagel (UM), Yanjie Sun (UM), and Munish Suri (CMU) over the summer of 1996 and Christopher Caruana (UM), Dai Kawano (UM), Brian Nakai (CMU) and Pradya Prempraneerach (CMU) over the summer of 1997.

Graduate student Jonathon Luntz (CMU) wrote the Simulink tutorials and contributed significantly in preparing the tutorials for web publication.

With further funding by MathWorks in 2011 and 2017, Prof. Bill Messner, Prof. Rick Hill (Detroit Mercy University), and Ph.D. Student J.D. Taylor (CMU), expanded the tutorials, completely redesigned the web interface, and updated all of the tutorials to reflect new functionality and tools available in the most recent version of the software.

Further funding from the MathWorks in 2019 supported Prof. Shuvra Das (DM) to develop the Simscape models included for each of the examples.

All the systems studied in this chapter are taken from previous website. This website is a valuable reference for anyone who wants to learn the process of designing a state-space controller for linear systems. The reader is encouraged to study other systems available there to learn more.

Author's Biography

FARZIN ASADI

Farzin Asadi received a B.Sc. in Electronics Engineering, M.Sc. in Control Engineering, and Ph.D. in Mechatronics Engineering. Currently, he is with the Department of Electrical and Electronics Engineering at the Maltepe University, Istanbul, Turkey. Farzin has published more than 35 international papers and 10 books. He is on the editorial board of 7 scientific journals as well. His research interests include switching converters, control theory, robust control of power electronics converters, and robotics.

Printed in the United States
by Baker & Taylor Publisher Services